Cave and Karst Systems of the World

Series Editor

James W. LaMoreaux, P. E. LaMoreaux and Associates, Tuscaloosa, AL, USA

This book series furthers the understanding of cave and karst related processes and facilitates the translation of current discipline-specific research to an interdisciplinary readership by dealing with specific cave or karst systems. Books in this series focus on a specific cave or karst system, on the cave or karst systems of a specific region, on a specific type of cave or karst system, or on any other perspective related to cave and karst systems of the world. The book series addresses a multidisciplinary audience involved in anthropology, archaeology, biology, chemistry, geography, geology, geomorphology, hydrogeology, paleontology, sedimentology, and all other disciplines related to speleology and karst terrains.

Nikolay Maksimovich • Olga Kadebskaya •
Olga Meshcheriakova

Sulfate Karst of Perm Region (Russia)

Springer

Nikolay Maksimovich
Institute of Natural Sciences
Perm State University
Perm, Russia

Olga Meshcheriakova
Perm State University
Perm, Russia

Olga Kadebskaya
Mining Institute of the Ural Branch
Perm, Russia

ISSN 2364-4591 ISSN 2364-4605 (electronic)
Cave and Karst Systems of the World
ISBN 978-3-031-18971-5 ISBN 978-3-031-18972-2 (eBook)
https://doi.org/10.1007/978-3-031-18972-2

Translation from the Russian language edition: "Сульфатный карст Пермского края" by Nikolay Maksimovich et al.,

This Springer imprint is published by the registered company Springer Nature Switzerland AG
The registered company address is: Gewerbestrasse 11, 6330 Cham, Switzerland

Dedicated to the outstanding karstologist Klara A. Gorbunova ...

Preface

The scientific work (monograph) provides a characteristic of sulfate karst, the features of its distribution and development in the Perm Region, based on their own research and generalization of data accumulated in recent years. An updated zoning of the territory is given, partly based on the use of GIS technologies.

A detailed description of areas with characteristics of geological and hydrogeological conditions for the development of karst and karst phenomena is given. A detailed description of a number of caves is given, information about specially protected natural areas associated with sulfate karst is provided, and new unique objects that need protection are proposed. The book describes the influence of karst on economic activity and the ecological situation.

The publication is intended for geologists, geographers, ecologists, karst scientists, speleologists, local historians, teachers, and students of specialties related to earth sciences, as well as a wide range of karst and cave enthusiasts.

Perm, Russia

Nikolay Maksimovich
Olga Kadebskaya
Olga Meshcheriakova

Acknowledgments

The authors are grateful to colleagues from the Perm State National Research University: E. O. Bezmaternykh, M. V. Bogomaz, O. A. Berezina, A. D. Demenev, E. V. Gubina, A. S. Kazantseva, V. T. Khmurchik, I. V. Kitaeva, A. V. Krasikov, Ya. R. Polyakova, S. V. Pyankov, A. M. Sedinin, Yu. N. Shutova, as well as L. N. Basharina, I. Yu. Gerasimova, O. O. Dodonova, and other cavers for their help in the preparation of this scientific work.

Introduction

The Perm Region is a kind of a model for most of the processes of karst development, despite the small area (1% of Russia), which is characterized by a wide variety of natural conditions. Tectonically, this region is located at the junction of the East European platform and the Central Ural uplift, connecting with the latter through the Cis-Ural foredeep and the forward fold zone of the Urals. This determines the presence of rocks of various ages and origins and a wide variety of minerals. About 30% of the territory is occupied by karst rocks: carbonates, sulfates, and salts.

The location of the Region within the temperate zone, its considerable length in the meridional direction, and the complex relief determine the presence of latitudinal and altitudinal zoning. On the territory of the Region there are zones of taiga, subtaiga, and mountain tundra. Almost the entire Region is the catchment area of the Kama River, the largest tributary of the Volga River. Within the territory of the Region, two large reservoirs have been created—Kama (1954) and Votkinskoye (1964).

The technogenic impact on the territory of karst development is also diverse. The Perm Region is an old industrial region with a long history of development, which began in the fifteenth century, when salt production emerged here. The first salt pans, built by the Stroganov merchants in the north of the region, served as a powerful incentive for the influx of the economically active population here.

Soon after the discovery of salt deposits, copper sandstone deposits were found, on the basis of which the first copper smelter was built in 1640. In the eighteenth century, 16 ore deposits were discovered here, with the development of which the emergence of the largest metallurgical centers began.

Smelting the metal required significant energy costs, which were compensated by the annealing of charcoal. By the end of the eighteenth century, charcoal was replaced by coal. Its deposits were discovered in 1786 within the Kizel Coal Basin.

1929 was marked by the discovery of the Permian oil. Some of the deposits are located in areas of sulfate karst development. It is currently being produced at 800 wells. In addition to oil, one of the world's largest deposits of potassium and magnesium salts, Verkhnekamskoye, is being developed. Chromium ores, gold, diamonds, ornamental stones, building materials, etc. are mined. The natural resources of the region are the basis for the development of chemistry and petrochemistry, power engineering, mechanical engineering, metalworking, forestry, pulp and paper, and woodworking and mining industries.

The level of resource availability and economic development made the Perm Region one of the most industrially developed regions of Russia, but at the same time it could not but have a significant impact on its natural and ecological potential and in some cases led to the technogenic activation of karst.

Sulfate rocks—gypsum and anhydrite—spread over 23 thousand km^2 (15%) of the territory of the Perm Region largely determine the originality of its natural conditions, primarily due to the development of karst processes, which condition a wide variety of surface and underground landforms and their inherent landscapes.

Sulfate karst, in comparison with the carbonate one, which is more widespread in the world, has a specificity associated, first of all, with the high solubility of gypsum and anhydrite, and, accordingly, the activity of karst processes, which must be taken into account especially in conditions of intense anthropogenic impact.

A distinctive feature of the Perm Region is the areas of joint occurrence of sulfate and salt-bearing rocks. Extensive wetlands are formed here, including the largest lakes in the Perm Region. The areas of development of sulfate rocks are characterized by extensive depressions, swamps, an abundance of karst lakes, dry lands, and disappearing and reappearing rivers. Groundwater and surface waters contain large amounts of calcium sulfate, which in some cases make them unsuitable for water supply and create difficulties for the local population.

A serious problem of sulfate karst regions is the development of sinkholes that damage buildings and structures, industrial enterprises, communications, and lead to the death of people.

Karst regions are characterized by increased water exchange, low protection of groundwater from pollution, which contributes to the development of environmental problems.

At the same time, gypsum and anhydrite are valuable mineral and construction resources. About 270 objects of sulfate rocks in 20 administrative districts are known in the Region.

Karst forms—caves, lakes, rivers, and others—are unique natural monuments, some of them have been declared as specially protected natural areas. The Kungur Ice Cave is the oldest tourist cave in Russia, and the world's longest underwater gypsum cave—The Orda Cave—is a world-famous center of underwater speleology.

This monograph is devoted to these and other unique features of the sulfate karst regions of the Perm Region.

The authors refer to regions of sulfate karst which includes both regions where only gypsum and anhydrite are karsted, and those where gypsum and anhydrite are karsted together with carbonate rocks and salts.

The last major work devoted to this problem, "Karst and Caves of the Perm Region" was published on the initiative of *K. A. Gorbunova* in 1992. The zoning of karst carried out by her has become generally recognized, and the monograph is widely cited in Russia and abroad.

When writing this monograph, the authors adhere to the principles of zoning and the structure of the description of karst phenomena proposed by *K. A. Gorbunova.*

Over the past almost 30 years, new data has appeared on sulfate karst and its manifestations in regions not covered by the previous zoning. During the work on the monograph, the GIS "Sulfate Karst of the Perm Region" was created, which contains information on the number of karst sinkholes within the regions since 1920, places of discharge of sulfate waters, the presence of sulfate rocks in outcrops, coordinates of caves, their morphometric parameters, and existing and the proposed protected natural areas with manifestations of gypsum karst. To summarize the information, the ArcGIS program was used. Based on this, some boundaries of the regions were clarified. The authors considered it possible to combine the karst sites identified by *K. A. Gorbunova* into a single Lysva Region, where sulfate rocks do not have continuous distribution, but the processes have common features.

The study of the karst of the Perm Region began in the 60s of the eighteenth century and is reflected in the works of such scientists as I. P. Falk, I. I. Lepekhin, and G. Benedict. In the nineteenth–early twentieth centuries, the study of karst was continued by Chekletsov, Ryshkovsky, Meyer, Platonov, I. F. Erdman, M. Kittary, R. I. Murchison, P. I. Krotov, I. S. Polyakov, E. S. Fedorov, S. I. Sergeev, A. A. Kruber, N. I. Karakash, A. A. Shtukenberg, and V. A. Varsanofieva. In the post-revolutionary period, significant contributions to research were made by G. Frederiks, N. P. Gerasimov and E. I. Tikhvinskaya, S. M. Orlyankin, S. Gurevich and E. A. Petrova, N. K. Tikhomirov, G. S. Burenin, V. P. Guldenbalk and M. S. Polozova, A. A. Turtsev, I. M. Pereslegin, and others.

The first brief global reviews specifically dealing with gypsum karst were provided by Maximovich (1962). Since then knowledge of gypsum karst, in terms of its morphological and hydrogeological peculiarities, development mechanisms, and geographical distribution, has increased dramatically. Recently, the global distribution of gypsum karst has been considered by Nicod (1976, 1993), Cooper and Calow (1998), Klimchouk et al. (1996), and many other authors (*Milanović et al. 2019*). Russian scientists such as G. A. Maksimovich, K. G. Butyrina, V. N. Dublyansky, E. P. Dorofeev, V. P. Zverev, V. S. Lukin, I. A. Pecherkin, A. V. Turyshev, P. I. Yakovenko, R. V. Yashchenko, and others dealt with the problem of sulfate karst in the middle and late twentieth century.

A special role in the study of sulfate karst in the Perm Region belongs to K. A. Gorbunova. She prepared the monographs "Features of gypsum karst" (1965), "Karst of gypsum of the USSR" (1977), "Morphology and hydrogeology of gypsum karst" (1979), and " Karst studies. Issues of typology and morphology of karst"(1985), which reflect the problems of sulfate karst in the Region.

In recent years, V. N. Andreychuk, V. G. Beltyukov, G. N. Dublyanskaya, D. R. Zolotarev, E. A. Ikonnikov, V. N. Kataev, Yu. A. Kilin, T. G. Kovaleva, V. P. Kostarev, N. V. Lavrova, I. I. Minkevich, N. N. Nazarov, S. V. Shcherbakov, as well as the authors of this monograph have contributed to the study of sulfate karst.

References

Cooper, A.H., & Calow, R.C. (1998). *Avoiding gypsum geohazards*: Guidance for planning and construction. British geological survey. Technical report WC/98/5 overseas geological series.

Gorbunova, K.A. (1965). *Features of gypsum karst*. Perm: Perm publishing house. http://nsi.psu.ru/cave/bibl/ogk.pdf. (in Russian).

Gorbunova, K.A. (1977). *Karst gypsum of the USSR*. Perm: Publishing house of Perm University. http://nsi.psu.ru/cave/bibl/kgu.pdf. (in Russian).

Gorbunova, K.A. (1979). *Morphology and hydrogeology of gypsum karst*. Perm: Perm University. http://nsi.psu.ru/cave/bibl/mggk.pdf. (in Russian).

Gorbunova, K.A. (1985). *Karst studies. Issues of typology and morphology of karst*. Perm: Perm University. http://nsi.psu.ru/cave/bibl/kvtimk.pdf. (in Russian).

Gorbunova, K.A., Andreychuk, V.N., Kostarev, V.P., Maksimovich, N.G. (1992). *Karst and caves of the Perm region*. Perm: Publishing house of Perm University. http://nsi.psu.ru/labs/gtp/stat/ng_0129.pdf. (in Russian).

Klimchouk, A., Lowe, D., Cooper, A.H., Sauro U. ed. (1997). Gypsum karst of the world. *International Journal of Speleology*, 5(1–2).

Milanović, P., Maksimovich, N., Meshcheriakova, O. (2019). Dams and Reservoirs in Evaporites. Springer, Cham.

Nicod, J. (1976). Karsts des gypses et des evaporates associees. *Annales de Geographie*, 417, 513–554.

Nicod, J. (1993). Resherches nouvells sur les karsts des gypses et des vaporates associees (Seconde partie geomorphologie, hydrologye at impact anthropique). *Karstologia*, 21, 15–30.

Contents

About the Authors

Nikolay Maksimovich began dealing with problems of karst and caves while studying at the Department of Engineering Geology and Environmental Protection of the M. V. Lomonosov Moscow State University (1973–1978).

In 1984 at Moscow State University Nikolay defended his Ph.D. thesis "Post-injection processes during plugging of gypsum-bearing carbonate rocks with silicate solutions (by the example of the Kama hydroelectric power station)". He went on to work at the Perm University in the laboratory of geology of technogenic processes which he organized. Since 1998, he has served as the Deputy Director for Research Work of the Institute of Natural Sciences the Perm State National Research University.

Along with research in the field of engineering geology, hydrogeology, ecology, technogenic geochemistry, a significant part of Nikolay Maksimovich's works is devoted to various issues of studying karst and caves. He considered theoretical issues of speleology and karst studies, in particular the role of karst in the evolution of the earth's crust. Together with Klara A. Gorbunova, the types of karst-forming environments on the territory of the USSR were identified, which are based on taking into account the geological factors and physical-geographical conditions that have developed during geological history (the ratio of heat and moisture). The regularities of the karst of the Urals and the Perm Region have been studied, including under conditions of anthropogenic impact, as well as on the banks of reservoirs. The results of the work were summarized in the collective monograph "Karst and Caves of the Perm Region" (1992).

Since the beginning of the XXI century, Nikolay Maksimovich began dealing with the problems of environmental pollution during oil production in karst regions, and also considers the engineering-geological and hydrogeological patterns of the development of karst, in particular, during the construction of nuclear power facilities in Belarus in the sites of chalk deposits.

Under his leadership, in 2007–2008, an assessment of the karst hazard and a forecast of possible changes in the hydrogeological situation during the construction and operation of the structures of the Sangtuda hydroelectric power station on the Vakhsh River (Tajikistan), at the base of which karst gypsum rocks lie, were made. Subsequently, based on the generalization of the world experience in the construction of dams on evaporites, in collaboration with Professor Peter Milanović of the University of Belgrade and Olga Meshcheriakova, the monograph "Dams and Reservoirs in Evaporites" was published at Springer (2019).

A significant part of Nikolay's work is related to the karst of the Kizel coal basin (Perm Region)—a territory of ecological disaster, as a result of the study of which methods were proposed for purifying and reducing the volume of acid mine water outflows.

Nikolay Maksimovich pays attention to the study of the mineralogy, hydrogeology, radioactivity, microbiology of caves, methods of their study, the use of caves as tourist sites. A number of works are devoted to the comprehensive study of the Kungur Ice and Orda caves.

Over the years of his scientific activity, Nikolay Maksimovich has published more than 570 scientific works in 25 countries of the world in 5 languages, including 24 monographs and textbooks such as "Karst and Caves of the Perm Region" (1992), "Kungur Ice Cave: The

Experience of Regime Observations" (2005), "Safety of Dams on Soluble Rock (The Kama hydroelectric power station as an example)" (2006), "Small Reservoirs: Ecology and Safety" (2012), "Orda Cave: Cognition" (2011, 2017). He is the author of chapters in books "Hypogene Karst Regions and Caves of the World" (2017, Springer), "Assessment, Restoration and Reclamation of Mining Influenced Soils" (2017, Elsevier). He is the author of 12 patents, mainly in the field of environmental protection. Since 2008, he has been the editor in chief of collection of scientific publications "The Caves" which was published since 1947.

Participant of major projects was related to solving geological and environmental problems of the Caucasus and the Urals in various regions of Russia, Kazakhstan, Tajikistan, and Uzbekistan. In 2016, he was the scientific leader of the expedition of the Center for Underwater Research of the Russian Geographical Society for the study of the Tserik-Kel karst lake in Kabardino-Balkaria.

To preserve the Orda Cave (Perm Region), the world's largest underwater cavity in gypsum, he proposed and carried out a justification for giving it the status of a specially protected natural area of regional significance.

He completed various international scientific trainings, such as "Geologically Active" (New Zealand), "Environmental Technologies in Karst Areas" (Slovakia), "Environmental technologies in the exploration and production of minerals" (Germany), "Classical Karst (Karst Underground Protection)" (Slovenia).

Nikolay Maksimovich is a member of the National group of International Association of Engineering Geology and the Environment (IAEG), International Geochemical Society, Institute of Karstology and Speleology, Russian Geographical Society, and Associate Member of International Show Caves Association. In the mid 1990s, he participated in the organization of the Italian-Russian Institute for Environ-mental Research and Education.

He was honored Ecologist of the Russian Federation, expert of the Russian Academy of Sciences.
e-mail: nmax@psu.ru

Olga Kadebskaya graduated in 1997 from the Faculty of Geography, Perm University, specializing in geography and ecology. In 2001, she entered the full-time postgraduate study of the State Institute of the Ural Branch of the Russian Academy of Sciences, where from the same time she has been working as the head of the Kungur laboratory. Since 2008, she has been an associate professor at the Department of Tourism, Faculty of Geography, Perm State National Research University.

In 2004 Olga Kadebskaya received a Ph.D. in Geography, having defended her thesis on the topic "Geoecological state of the Kungurskaya Ice Cave and the adjacent territory, its protection and rational use."

The main directions of scientific activity: the study of the patterns of karst development in different types of economic development of karst regions, as well as the study of the caves of the Urals and Siberia.

Olga Kadebskaya took part in the grants of the Russian Foundation for Basic Research "Development of a methodology for georadar studies of underground ice in caves" (2011–2013), "Development of a method for identifying karst cavities and assessing the dynamics of their development based on local seismological monitoring data" (2007–2009), "Investigation of the formation of isotopic and chemical composition of natural waters in the Perm region", "Migration of the permafrost border on the Europe-Asia border in the Pleistocene" and "Pleistocene permafrost boundary shifts at the Europe-Asia border" (Austria).

Conducted a detailed description of the specially protected natural areas of the Perm Region associated with the manifestation of karst. She took part in projects to identify unique objects of the geological heritage of the Urals and the Perm Region and made proposals for their protection. Conducted a justification for granting the status of a specially protected natural area "Kamenny Gorod".

In 2017 Olga Kadebskaya defended her doctoral dissertation on the topic "Mineral and geochemical indicators of natural processes in underground karst landscapes of the Urals."

She is the scientific secretary of the editorial board of the annual collection of scientific articles "Caves", a member of the Russian Geographical Society, of the scientific committee of the international association of excursion caves of the world, of the scientific committee of the international karstological school, as well as the representative of Russia at the International Academy of Research.

Author of 300 scientific works, including 8 monographs. She obtained two certificates of state registration of the database.

Personal scholarship holder of the Perm Region (2004), laureate of the Perm University in the category "Earth Sciences" for the best research work (2018).
e-mail: icecave@bk.ru

Olga Meshcheriakova In 2007 Olga Meshcheriakova graduated from the magistracy of the Perm State National Research University with a degree in Geology. In 2014 she defended her Ph.D. thesis on "Protection of gypsum deposits from oil pollution in the karst areas of the Perm Region" with the award of the degree of candidate of technical sciences.

Since 2007 Olga Meshcheriakova has been working as Deputy Dean of the Geological Faculty of Perm University. Associate Professor of the Department of Dynamic Geology and Hydrogeology, Senior Researcher of the Laboratory of Geology of Technogenic Processes of the Institute of Natural Sciences of Perm State National Research University.

The area of scientific interests of Olga Meshcheryakova relates to issues of geoecology, karst studies, hydrogeology. The main research topic is oil pollution of karst regions.

The results of research by Olga Meshcheryakova were presented at international conferences in Portugal (2020), Croatia, Serbia, Kyrgyzstan (2019), Austria, Bosnia and Herzegovina (2018), Japan (2016), New Zealand (2010), China, Egypt (2009), Italy (2006), Russia (2008–2022). She takes part in research and contractual work. She is the head and executor of grants of the Russian Foundation for Basic Research, Federal Target Program, and others.

Author of more than 100 scientific works, including 3 monographs and one textbook. Her monograph "Dams and Reservoirs in Evaporites" was published at Springer (2019) in collaboration with Professor Peter Milanović of the University of Belgrade and Nikolay Maksimovich. She is a member of the editorial board of the collection of scientific articles "Caves".

She completed various international scientific trainings, such as "Geologically Active" (New Zealand), "Environmental technologies in the exploration and production of minerals" (Germany), "Visual KARSYS: a web-tool for modelling karst aquifers in 3D" (Switzerland), "Environmental Technologies in Karst Areas" (Serbia).

Olga Meshcheriakova is a member of the National group of International Association of Engineering Geology and the Environment (IAEG), the Karst Committee of the International Association of Hydrogeologists (IAH), International Geochemical Society, Institute of Karstology and Speleology, Russian Geographical Society, and Associate Member of International Show Caves Association. Participant of International symposia, conferences on geotechnical challenges, karst and hydrogeology, environmental pollution and environmental management.

Laureate of the competition for the best research work in the field of earth sciences among young scientists of Perm University (2014, 2017); winner of a travel grant of the Endowment Fund of the Perm University (2016).
e-mail: olgam.psu@gmail.com

1 Features of Sulfate Karst

Photo by M. Maksimovich

On the global scale, surface outcrops of gypsiferous strata appear quite limited. This apparent scarcity can be explained by the relatively low resistance of gypsum (a soft rock) to all types of denudation processes, rather than being a limited occurrence of sulfate rock deposits.

Sulfate rocks cover vast areas of the earth's surface (Klimchouk et al. 1997): according to various estimates, from 7 million km^2 (Maksimovich 1963) to 60 million km^2 (Ford and Williams 1989, 2007).

Karst processes operate extensively in intra-stratal settings beneath various types of cover beds, particularly where the gypsum/anhydrite beds occur within at least the upper few hundred meters of a stratigraphic sequence. Taking this into account, gypsum karst appears to be a much more widely developed phenomenon than commonly believed.

The largest areas of sulfate rocks are found in the northern hemisphere, especially in the USA, Canada, Russia, China, Iran, Spain, and Great Britain (Fig. 1.1), and are developed from arctic latitudes to hot arid or humid tropics, from the very lowlands to the highlands (Chen et al. 2017; Cooper and Gutiérrez 2013; Cooper and Saunders 2002; Cucchi and Zini 2003).

In the USA, gypsum deposits are present in 32 states, and they underlay about 35–40% of the regions where Precambrian through Quaternary strata are in outcrop (Dean and Johnson 1989; Quinlan 1986).

N. Maksimovich et al., *Sulfate Karst of Perm Region (Russia)*, Cave and Karst Systems of the World,
https://doi.org/10.1007/978-3-031-18972-2_1

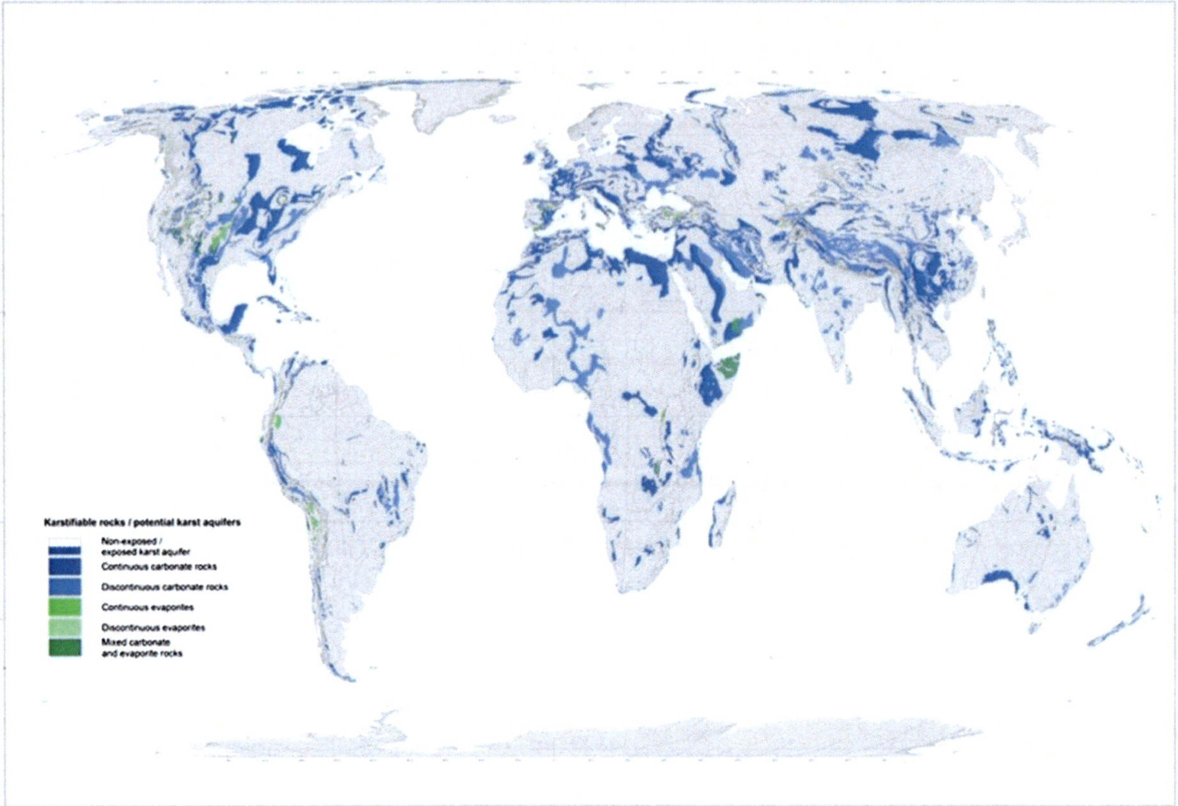

Fig. 1.1 World distribution of karstifiable rocks (Chen et al. 2017)

According to Lu and Cooper (1996), China has the largest gypsum resources in the world. They range in age from Precambrian to Quaternary, and their genesis includes marine, lacustrine, thermal (volcanic and metasomatic), metamorphic, and secondary origin.

In Iran, there are large areas of rock formations containing evaporites, mostly gypsum and salt. The total outcrop of the gypsum formations is about 80,000 km 2.5% of Iran's land area (Raeisi et al. 2013).

According to Mancebo Piqueras et al. (2011), Spain has the highest proportion of gypsiferous rock outcrops in the world. Broadly, gypsiferous geological formations occupy 21% of the total surface area of the nation, with some 9% being massive evaporite formations.

Saline and gypsum soils occupy vast territories of Australia, Asia (Afghanistan, India, Iraq, Iran, China, Mongolia, Syria, Pakistan), Africa (Algeria, Egypt—especially in the Nile Delta, Libya), North and South America (Argentina, Canada, Mexico, USA, Chile, Peru), and Europe (Greece, Italy, Spain, Romania, France) (Petrukhin 1993). According to Chokhonelidze (1957), the total area of saline and gypsum soils is about 750 thousand km^2. According to estimates by Nafie (1989), gypsum soils occupy 724 thousand km^2 of the continental surface.

In Russia and the surrounding former USSR states, Gorbunova (1977, 1981, 1993a, b; Gorbunova and Maksimovich 1988) estimated the extent of sulfate rocks to be about 5 million km^2 (Fig. 1.2).

Soils containing salts, including gypsum, are widespread throughout the territories of many countries of the world. The origin of salts in soils is associated with the processes of dissolution and chemical weathering of rocks, which cause the transition of their constituent minerals into salt solutions. Salt accumulation in soils is typical for semi-arid and arid regions with a dry climate and a negative moisture balance in soils, where the annual amount of atmospheric precipitation does not exceed 25 mm/year (Askhor 2003; Petrukhin 1993).

Solubility of gypsum and anhydrite. The solubility of sulfate rocks and, accordingly, the activity of karst processes depend both on their chemical composition and structure and on external factors: pressure, temperature changes, pH, redox potential, chemical composition and hydrodynamic characteristics of groundwater, as well as microbiological processes.

Let us consider the influence of the factors most significant for the Perm Region.

Influence of the composition of groundwater. Sulfate rocks on the territory of the Perm Region are paragenetically

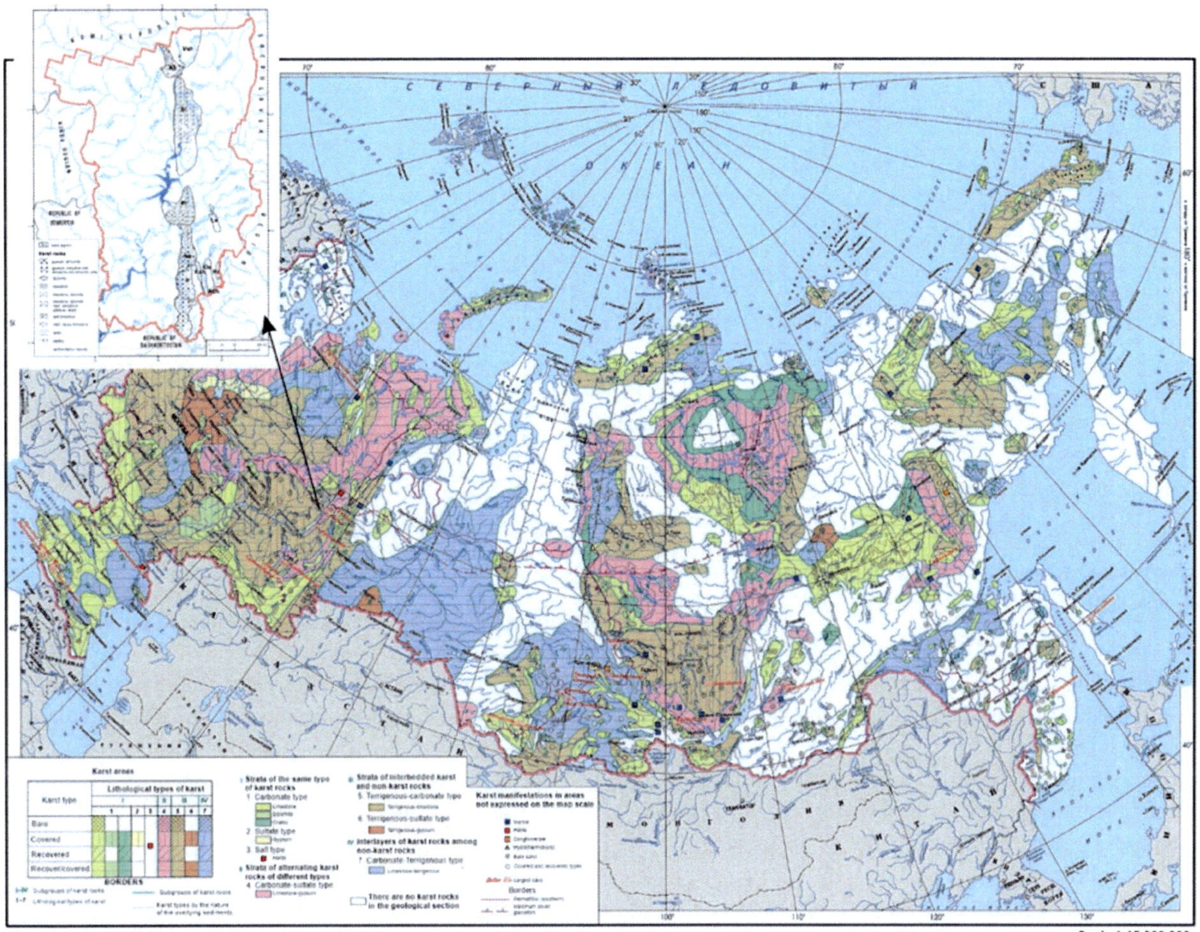

Fig. 1.2 Map of karst hazard of the territory of Russia (https://geographyofrussia.com/karstovye-peshhery-rossii/)

associated with salt-bearing rocks, mainly with halite and sylvinite. The concentration of sodium chloride in groundwater in such massifs varies widely—up to saturated solutions (about 360 g/dm^3).

Experiments by Shternina (1949) show that the solubility of gypsum in distilled water at a temperature of 25 °C is 2.1 g/dm^3. The usual concentration of calcium sulfate in groundwater is 2–2.5 g/dm^3, and at a NaCl concentration of about 130–140 g/dm^3, the solubility of gypsum reaches a maximum value of 7.48 g/dm^3. With an increase in the concentration of NaCl to 293 g/dm^3, it decreases to 5.19 g/dm^3 (Fig. 1.3).

This feature of calcium sulfates is also confirmed by the experiments of E. Bock (Fig. 1.4; Bock 1961).

A significant increase in the solubility of gypsum in areas of joint occurrence with salts can affect the activity of both natural and technogenic karst processes. It is known that during the development of salt deposits, powerful technogenic flows with a high content of sodium chlorides arise.

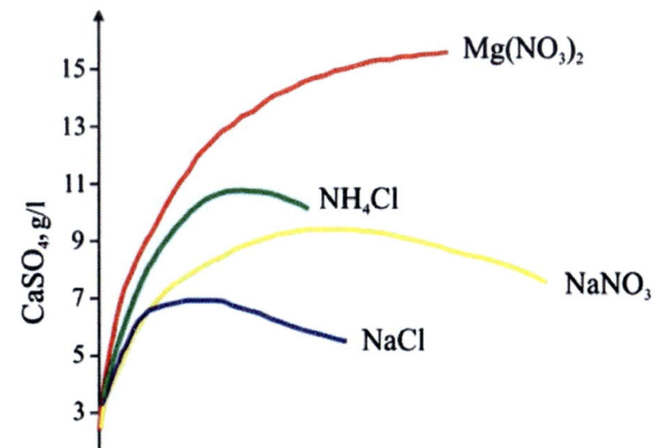

Fig. 1.3 Solubility of calcium sulfate in aqueous solutions of salts at 25 °C (Shternina 1949)

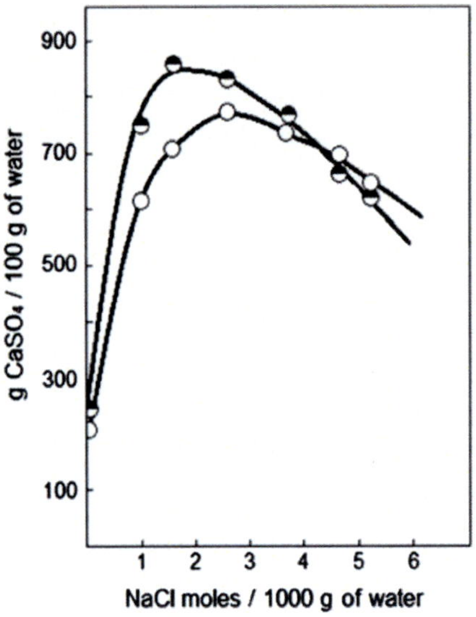

Fig. 1.4 Experimentally modified ionic effect of NaCl on gypsum and anhydrite at 25 °C (Bock 1961)

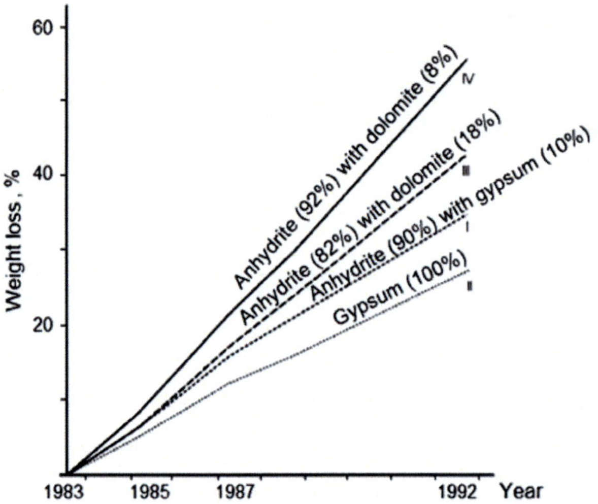

Fig. 1.5 Weight loss of samples (in%) for observation period I, II, III, IV—sample numbers (Gorbunova et al. 1986, 1993a)

The solubility of gypsum depending on the concentration of certain salts is shown in Fig. 1.3. Attention is drawn to the increase in the solubility of gypsum in the presence of nitrogen compounds. Such compounds are formed in significant quantities as a result of the activities of agricultural enterprises, which are especially numerous in the southern part of the Perm Region.

In solutions representing mixtures of various salts, the solubility of gypsum obeys more complex laws, which were investigated in the works of Zverev (1967), Ponizovsky and Pachepsky (1979), and others.

Changes in temperature and pressure do not significantly affect the solubility of gypsum in this region. Fluctuations in the temperature of groundwater and infiltration waters are insignificant, with the exception of the uppermost part of the section. The pressure in the zone of active water exchange does not change significantly and, therefore, practically does not affect the solubility of calcium sulfate (Gorbunova 1977).

Various authors note the dependence of gypsum solubility on its structure and composition (Brune 1965; Christoffersen and Christoffersen 1976; Rodionov 1958; Skvortsov 1955; Sokolov 1962), but, given the natural heterogeneity of gypsum, these data are difficult to use to assess the effect of this factor for each specific case.

The factors listed above highlight the difficulties that have to be faced when studying rock dissolution. Completely objective information can be obtained by carrying out dissolution experiments under natural conditions. In this case, there is no need to reproduce in the laboratory the entire variety of external conditions and their seasonal changes. Let us give an example of an experiment to study the solubility of gypsum and anhydrite in the conditions of the Kungur Ice Cave.

In order to study the process of dissolution of rocks by karst waters close to their saturation with calcium sulfate, a natural experiment was carried out (Gorbunova et al. 1986, 1993a). Samples of gypsum-anhydrite rocks of the Irenian horizon of the Kungurian stage, about 5 × 5 × 1.5 cm in size, were suspended on a nylon fishing line in the lake of the Dlinny grotto at a depth of 0.5–0.7 m. The mineral composition of the samples was determined by X-ray (Fig. 1.5).

The experiment lasted over 10 years. During the observation period, the weight of all samples decreased by 27.49–50.95% (Fig. 1.5). Dissolution took place in sulfate-carbonate water, the salinity of which varied from 2.0 to 2.7 g/dm^3, the content of sulfate ion—from 1.3 to 1.7 g/dm^3. With an increase in mineralization, calcite formations appeared on the surface of the lake in the form of spots or a continuous film, including rare gypsum crystals. The water temperature was from 3.0 to 4.4 °C and the air temperature was from 2.9 to 4.4 °C.

According to the data of routine observations in the cave, there are periods of spring water level rise caused by floods on the Sylva River during snowmelt, and autumn floods associated with the infiltration of summer and autumn precipitation into the massif. The spring periods correspond to a decrease in salinity and the content of sulfate ions as a result of mixing with less saline river waters, in winter—a slight increase in salinity due to the inflow of fissure-karst waters from the deep parts of the massif. In the initial period of 2.3 years, the weight of all samples decreased by 9–13% (Fig. 1.5). The surface of gypsum specimens (II) was covered with a powdery coating upon drying. Anhydrite with gypsum (I) was complicated by microcarriages. On samples of anhydrite with dolomite (III, IV), caverns with a depth of 2–5 mm and 1–3 cm in diameter were noted, and larger cavities were found on sample IV.

In subsequent periods, the surface relief of the samples became more contrasting. At low temperatures, the relative rate of dissolution of anhydrite with dolomite (III and IV) in sulfate waters is 1.7–2.4 times higher than that of gypsum (II). The dissolution processes of gypsum and anhydrite are seasonal in nature and are controlled by changes in the magnitude of the deficiency in water saturation with calcium sulfate.

Studies have shown that the process of dissolution of gypsum and anhydrite has complex patterns. The relative rate of dissolution changes over time depending on many factors, in particular, on the nature of the interaction of the rock with water, the composition and mineralization of the latter, the mineral composition of the rock, and the temperature regime. The results of the experiment allow us to conclude that the dissolution processes, to a certain extent, actively proceed under conditions close to the saturation of waters with calcium sulfate (Gorbunova et al. 1986, 1993a, b).

References

Askhor, Kh.M. (2003). *Accounting for the dissolution process of gypsum contained in soils when calculating the deformation of the foundations of buildings*: PhD thesis. St. Petersburg (in Russian).

Bock, E. (1961). On the solubility of anhydrous calcium sulphate and of gypsum in concentrated solutions of sodium chloride at 25 °C, 30 °C, 40 °C, and 50 °C. *Canadian Journal of Chemistry*, 39, 1746–1751.

Brune, G. (1965). Anhidrite and gypsum problems in engineering geology. *Engineering Geology* (Sacramento). Publ. 2: 26–33.

Chen, Z., Auler A., Bakalowicz M., Drew D., Griger F., Hartmann J., Jiang G., Moosdorf N., Richts A., Stevanovic Z., Veni G., Goldscheider, N. (2017). The World Karst Aquifer Mapping Project – Concept, Mapping Procedure and Map of Europe. *Hydrogeology Journal*, 25, 771–785.

Chokhonelidze, G.I. (1957). On the issue of assessing the physical and technical properties of saline soils used in hydraulic structures. *Trudy GruzNIIP*. 18(19), 23–43. (in Russian)

Cooper, A.H., & Gutiérrez, F. (2013). Dealing with gypsum karst problems: hazards, environmental issues and planning. In Treatise on Geomorphology. *Karst Geomorphology*, vol. 6, Frumkin, A. (ed.), 451–462.

Cooper, A.H., & Saunders, J.M. (2002). Road and bridge construction across gypsum karst in England. *Engineering Geology*, 65(2–3), 217–223.

Christoffersen, J., & Christoffersen, M.R. (1976). The kinetics of dissolution of calcium sulphate dihydrate in water. *Growth*, 35(1), 79–88.

Cucchi, F., & Zini, L. (2003). Gypsum karst of Zagros mountains (IR Iran). *Acta carsologica*, 32(1).

Dean, W.E., & Johnson, K.S. (eds). (1989). Anhydrite deposits of the United States and characteristics of anhydrite important for storage of radioactive wastes, 132. U.S. Geological Survey Professional Paper.

Ford, D., & Williams, P. (1989). *Karst geomorphology and hydrogeology*. London: Unwin Hyman.

Ford, D.C., & Williams, P. (2007). *Karst Hydrogeology and Geomorphology*. John Wiley, Chichester.

Gorbunova, K.A. (1977). *Gypsum Karst of the USSR*. Perm: Publishing house of Perm University. http://nsi.psu.ru/cave/bibl/kgu.pdf. (in Russian).

Gorbunova, K.A. (1981). Gypsum-Anhydrite Karst on the Territory of the USSR. *Proceedings of the Eighth International Congress of Speleology*. Americus. I–II, 778.

Gorbunova, K.A., Dorofeev, E.P., Minkevich, I.I. (1993a). Experimental study of the solubility of sulfate rocks by groundwater in the Kungurskaya cave. *Caves (Peshchery)*, 23(24), 140–149. http://nsi.psu.ru/cave/vipuski/v23(24).pdf. (in Russian).

Gorbunova, K.A., Dublyanskaya, G.N., Dublyansky, V.N., Klimchouk A.B. (1993b). Karst Map of Former USSR. Geology, Climate, Hydrology and Karst Formation. Project IGCP 299: Newsletter. Guilin. 30–32.

Gorbunova, K.A., & Maksimovich, N.G. (1988). Types of conditions for karst formation in the USSR. *Soviet Engineering Geology (Inzhenernaya Geologiya)*, 4, 75–81.

Klimchouk, A., Lowe D., Cooper A.H., Sauro U. ed. (1997). Gypsum karst of the world. *International Journal of Speleology*, 5 (1–2).

Maksimovich, G.A. (1963). Osnovy karstovedeniya. *Fundamentals of karst studies. Issues of karst morphology, speleology and hydrogeology of karst*. Perm: Perm publishing house. http://nsi.psu.ru/cave/bibl/Osnovy%20karstovedeniya%20I.pdf. (in Russian).

Nafie, F.A. (1989). *The properties of highly gypsiferous soils and their significance for land management*. London. PhD thesis.

Ponizovsky, A.A., & Pachepsky, Ya.A. (1979). Determination of the solubility of gypsum in a five-component water-salt system Ca, Mg, Na, Cl, SO_4-H_2O at 25 °C by the calculation method. *Journal of Inorganic Chemistry*, 24(1), 161–167. (in Russian).

Petrukhin, V.P. (1993). *Regularities of deformation and calculation of bases composed of gypsum silty-clayey groups*: Dr. Sci. thesis. Moscow. (in Russian).

Piqueras M., Sanchez Perez E., Menendez-Pidal, I. (2011). Water seepage beneath dams on soluble evaporate deposits: a laboratory and field study (Caspe dam, Spain). *Bulletin of Engineering Geology and the Environment Springer-Verlag*.

Raeisi, E., Zare M., Aghdam, J.A. (2013). Hydrogeology of gypsum formations in Iran. *Journal of Cave and Karst Studies (Iran)*, 75, 68–80.

Rodionov, N.V. (1958). *Engineering-geological research in karst regions*. Moscow: Gosgeoltekhizdat. (in Russian).

Shternina, E.B. (1949). Solubility of gypsum in aqueous solutions of salts. *News of the sector of physical and chemical analysis of the Institute of General and Inorganic Chemistry of the Academy of Sciences of the USSR*, 17. (in Russian).

Skvortsov, G.G. (1955). On the rate of development of karst in gypsum. *Issues of studying groundwater and engineering-geological processes*. Moscow: Academy of Sciences of the USSR. (in Russian).

Sokolov, D.S. (1962). *Basic conditions for the development of karst*. Moscow: Gosgeoltekhizdat. (in Russian).

Quinlan, J.F., Smith, R.A., Johnson, K.S. (1986). Gypsum karst and salt karst of the United States of America. *Atti symposio internazionale sul carsismo nelle evaporiti. Le Grote d-Italia*, 4(13), 73–92.

Zverev, V.P. (1967). *Hydrochemical studies of the gypsum-groundwater system*. Moscow: Publishing house "Nauka". (in Russian).

2 Conditions for the Development of Sulfate Karst in the Perm Region

Photo by M. Maksimovich

2.1 Physical and Geographical Conditions

Karst rocks—limestone, dolomite, gypsum, anhydrite, and rock salt—occupy a significant part of the Perm Region. The area of distribution of sulfate karst crosses in a wide strip almost the entire territory of the region in the submeridional direction. The length of the strip is about 600 km and width up to 117 km, as a result of which the conditions for the development of karst are influenced by a variety of natural and climatic factors.

The Perm Region is located at the junction of two largest geological systems: the East European platform and the Ural Hercynian fold region. To the east, the platform is replaced by a zone of the Cis-Ural foredeep, passing into the West Ural fold zone and further into the Central Ural uplift (Engineering geology … 1978).

On the East European platform and in the Cis-Ural trough, flat and low-lying relief prevails. The eastern part of the region is located on the western slope of the Ural mountainous country. The hills and ridges of the western slope of the Urals, the medium–high mountains of the Northern Urals, and the low mountains of the Middle Urals rise here. The valley of the Kama River, which is the western boundary of the development of karst, is confined to the lowered central part of the Perm Region.

The area of distribution of sulfate karst is confined to the Cis-Ural foredeep, enters the extreme eastern part of the East

N. Maksimovich et al., *Sulfate Karst of Perm Region (Russia)*, Cave and Karst Systems of the World,
https://doi.org/10.1007/978-3-031-18972-2_2

European platform, and the extreme western part of the Ural fold region.

Within the Cis-Ural foredeep, three geomorphological regions are distinguished: the Cis-Ural Depression, the Cis-Ural Hilly Plain, and the Ufimian Plateau with the Sylvensky Ridge.

The Ural Depression plays an important role in the morphology of the river network. The presence of a vast low-lying plain with poor drainage has created favorable conditions here for excessive moisture and strong waterlogging.

The relief of the Ural Plain is quite hilly, and the size and orientation of the hills are very diverse. River erosion played the main role in the formation of the relief. Karst landforms are widely developed within the plain, in particular, sinkholes, some of which are filled with water and turned into lakes.

The relief of the Ufimian Plateau is flat and hilly in places. Its erosional dissection, although deep, is rare; in some places, it is connected with the karst process. The northern section of the Ufimian Plateau, oriented along the valley of the Sylva River, is called the Sylvensky Ridge. The heights on the Ufimian Plateau are usually 300–350 m. In the northern direction, the heights decrease and in most parts of the Sylvensky ridge do not exceed 280–300 m.

Within the areas of development of sulfate karst, flat and low-lying relief prevails, significantly indented by valleys of large and small rivers, streams, ravines; spread are smoothed relief forms. The root slopes of large rivers, such as the Chusovaya, the Sylva, and others, often have steep slopes, especially in the areas where karst rocks emerge on the surface. There is a general increase in surface elevations from west to east, up to the western slopes of the Ural Mountains.

The territory of the Perm Region is distinguished by a wide variety of landforms, which owe their origin to the action of three main factors: erosion, karst, and the latest tectonic movements. Karst is a significant relief-forming factor. An example is the depressions along the slopes of the Ufimian Plateau, formed as a result of the complete destruction of gypsum layers.

The nature of the relief, and, first of all, the steepness of the surface, its dissection by the gully-ravine network, and elements of the microrelief significantly affect the distribution of surface runoff and infiltration, thereby influencing the development of karst both in the aeration zone and in deeper horizons. The formation of the modern relief, including under the influence of karst processes, continues at the present time (Yakovenko 1969).

The climate of the areas where sulfate karst is distributed belongs to the continental temperate type and is formed mainly under the influence of the southwestern and western air currents coming from the Atlantic. The Ural Mountains located to the east have a significant influence on the regional climate, which trap moist air masses, causing an increased amount of precipitation. To a lesser extent, the local climate is affected by air currents breaking through from the east, from Siberia, through the lower part of the Ural Mountains, as well as by currents coming from the north and bringing severe frosts.

In general, the climate is characterized by long cold winters and short, moderately warm summers. The role of individual climate elements in karst formation is different. The most important are temperature, precipitation, and snow cover. They often play a decisive role, for example, in the frequency of occurrence of karst sinkholes (Yakovenko 1969). Precipitation and air temperature significantly affect the formation of natural waters, their composition, and water regime. Due to the considerable length of the region under consideration from north to south, the mean annual values of climatic indicators are somewhat different.

The temperature regime affects the duration of the snow cover, the time of the onset of the thaw, and the freezing of the soil. All this determines the conditions for moisture seepage and affects the development of karst. In addition, air temperature determines the characteristics of precipitation.

One of the most important temperature indicators is the average annual temperature, its low values contribute to the slow melting of snow and cause low evaporation, thus increasing water infiltration (Yakovenko 1969). The territory of the Perm Region is characterized by positive average annual temperatures. From the northeast to the southwest of the region, an increase in the temperature regime is noted.

Average annual temperatures within the study area vary from 0.1 to 2.8 °C, with the exception of a small area in the foothills of the Northern Urals, where the average annual temperature is minus 0.7 °C.

The coldest month of the year is January with an average monthly temperature of minus 14–minus 18 °C (Fig. 2.1). The absolute minimum of temperatures varies from minus 42.6 °C in the central part of the study area to minus 52.3 °C in its northern and southern outskirts. In winter, on hills, slopes, and mountain tops, the average January temperature is 1–3 °C higher than in intermountain valleys.

The warmest month of the year is July, the average monthly temperature of which ranges from 14 to 18 °C (Fig. 2.1). The absolute maximum temperature ranges from 34.1 to 37.2 °C.

Climatic winter (a period with an average daily air temperature below 0 °C) lasts on average from 164 to 184 days.

Spring frosts on average end in late May–early June, but in the northern part, which has a more severe climate, they can be observed until the end of June. The first autumn frosts begin on average at the very end of August–mid-September, in the northern part they can begin as early as August. The frost-free period lasts on average 82–111 days.

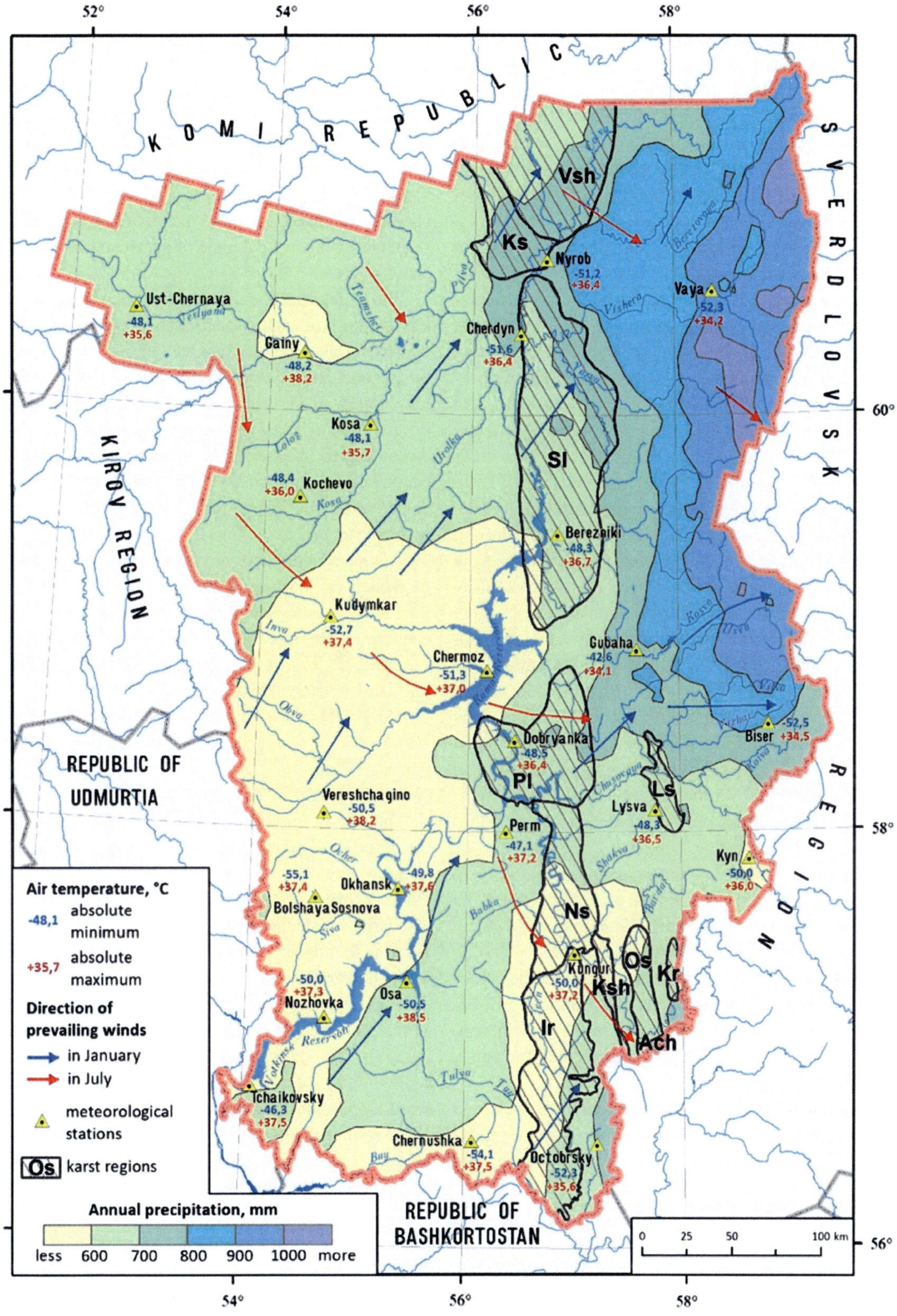

Fig. 2.1 Distribution of average temperatures, winds, and precipitation in areas of sulfate karst development in the Perm Region (Gorbunova et al. 1992; Maksimovich and Pyankov 2018). *Note* Regions of sulfate karst: Ks—Ksenofontovsky, Vsh—Visherksky, Sl—Solikamsky, Pl—Polaznensky, Ls—Lysvensky, Ns—Nizhnesylvinsky, Ksh—Kishertsky, Os—Osintsevsky, Ach—Achitsky, Kr—Kordonsky, Ir—Irensky

Precipitation regime. Precipitation is distributed very unevenly over the territory of the karst regions of the Perm Region. The influence of the barrier effect of the Ural Mountains on the nature of humidification is most pronounced in its northeastern margin of the study area, where on the western (windward) slopes the average annual precipitation is 900–1000 mm, rarely more than 1000 mm. As we move from northeast to southwest, the amount of precipitation decreases to 600 mm or less.

For the development of karst, not only the annual amount of precipitation is important, but also their distribution throughout the year. Within the territory under consideration, most of the annual precipitation will fall in the warm season (April–October), amounting to about 400–650 mm. The amount of precipitation in the cold period (November–March) is on average 150–350 mm. The maximum amount of precipitation (more than 600 mm in the warm season and more than 300 mm in the cold) falls on the western windward slopes of the ridges of the Northern Urals (Fig. 2.1). In the annual course of precipitation, the minimum usually falls in February–March (17–53 mm/month). A sharp increase in precipitation occurs in April–June, and in July–August the maximum precipitation values in the year are observed (71–99 mm). In autumn, the amount of precipitation, as a rule, decreases, but not significantly. Significantly less precipitation becomes from November. Analysis of the annual distribution of precipitation shows that in summer and spring, during the seasons of the highest precipitation, the most favorable conditions are formed for the development of karst processes. The dissolving ability of sediments increases with increasing temperature. According to long-term observations, the daily maximum precipitation varies from 53 to 114 mm.

The total amount of precipitation falling per year is spent on surface, slope runoff, evaporation, and infiltration. The annual evaporation is 320–340 mm (Efimova and Zubenyuk 1966). In the area of distribution of sulfate karst, infiltration and inflation exceed the slope runoff. Most of the atmospheric precipitation is spent on the restoration of groundwater resources. The underground runoff coefficient, expressed as a percentage of the annual precipitation, here is 20–22%.

The snow cover, having a high reflectivity, low heat, and gas conductivity, has a significant impact on the entire complex of natural conditions in winter in the temperate zone.

The role of snow cover in the processes of karst formation is that the melt waters formed in the spring are directly involved in the dissolution of sulfate rocks and, having low thermal conductivity, protect the soil from freezing, increasing its infiltration capacity. The period of snow melting is the most favorable for the development of karst (Yakovenko 1969).

The height and duration of the occurrence of the snow cover affect the freezing capacity of the soil, the characteristics of moisture accumulation, and the nature of weather conditions, which, in turn, determine the intensity of karst formation. During the winter, a large amount of water accumulates in the snow cover. Melted water is the main source of nutrition for surface and groundwater. The activation of karst processes is greatly influenced by the chemical composition of the moisture contained in the snow, which determines the degree of aggressiveness of the melted snow water.

Within regions of sulfate karst, stable snow cover, as one of the kinds of water accumulation on the earth's surface, usually forms closer to mid-October, its destruction occurs in the second half of April–early May. In some years, snow may appear as early as September. The latest dates for the destruction of the snow cover are dated May 7–18. According to long-term observations, the duration of snow occurrence is on average 160–200 days. In some small areas, snow can linger for more than 200 days and melt earlier than 160 days (Fig. 2.2).

The maximum depth of snow cover is formed by the end of winter–early spring (usually in March). For the considered territory, as well as for the territory of the entire Perm Region, an increase in the height of the snow cover and the water supply in it in the direction from the southwest to the northeast is characteristic (Fig. 2.2).

The height of the snow cover here varies on average from 60 to 100 mm, increasing on the western slopes of the Northern Urals to 120 mm and more. In winters with little snow, the height of the snow cover is 35–57 mm.

The maximum water content in the snow cover is formed by the end of March, its average value varies within the range of 171–268 mm. In some snowy winters, water content can exceed 400 mm (Scientific and applied… 1990). The largest water content, concentrated in the snow cover, is noted in the forests of the north of the study area, growing on the windward slopes.

Wind regime. The influence of wind on karst processes is indirect and noticeably less than precipitation and temperature. The main role of the wind is reduced to the redistribution of snow. Usually, the wind moves it to leeward slopes and to lower relief, including sinkholes, where large volumes of snow accumulate, and when it melts, karst processes are intensified. Snow in the sinkholes lasts much longer than on the surrounding surface of the earth (Fig. 2.3).

In the described territory, the prevailing winds are southern and southwestern. In the warm period, the northern, northwestern, and western transport are more often observed in the cold period, the southwestern, and western ones.

Thus, a fairly large amount of precipitation falls in the Perm Region, which contributes to the development of karst.

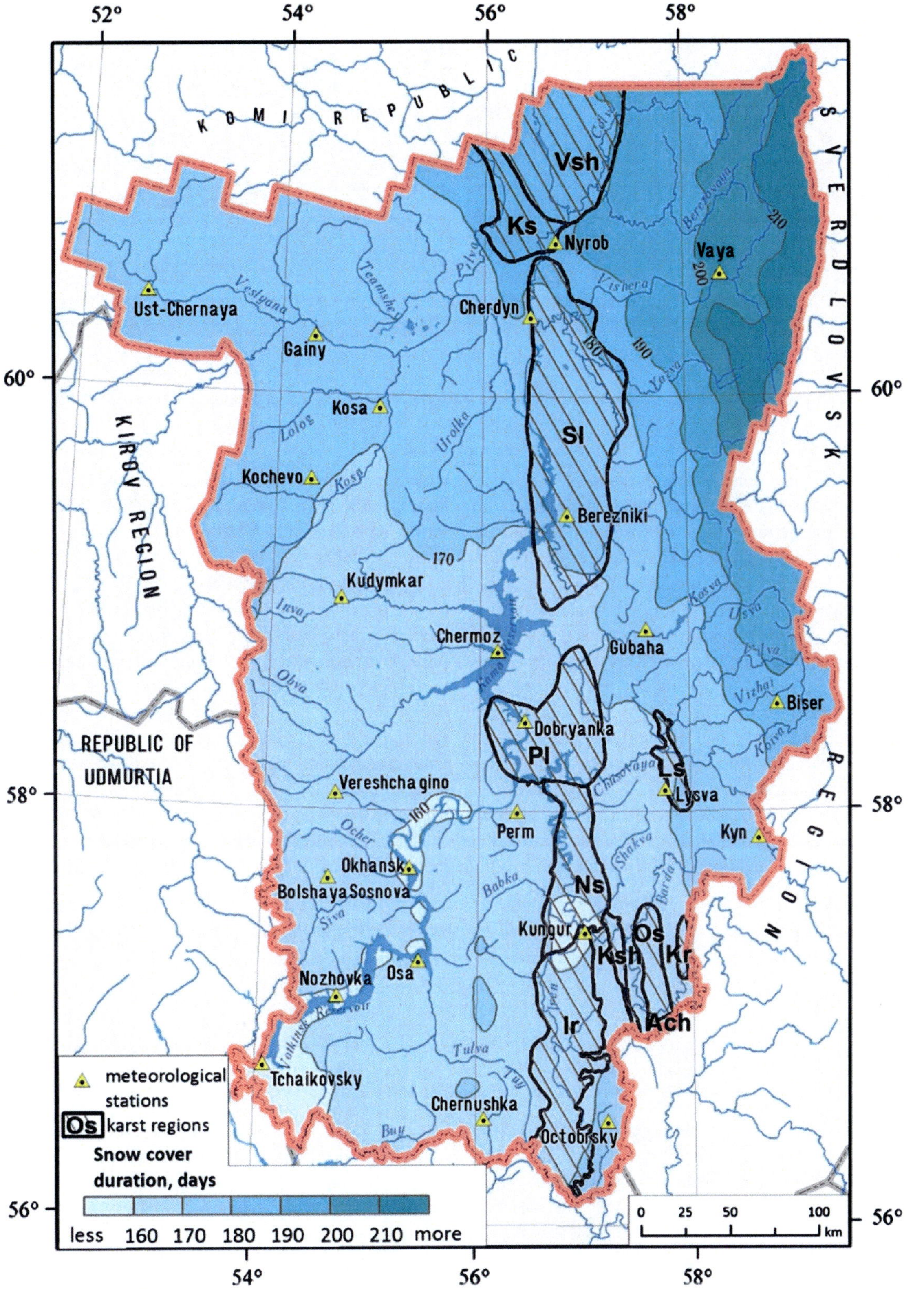

Fig. 2.2 Average duration of snow cover in regions of sulfate karst development in the Perm Region (Gorbunova et al. 1992; Maksimovich and Pyankov 2018). *Note* Regions of sulfate karst: Ks—Ksenofontovsky, Vsh—Visherksky, Sl—Solikamsky, Pl—Polaznensky, Ls—Lysvensky, Ns—Nizhnesylvinsky, Ksh—Kishertsky, Os—Osintsevsky, Ach—Achitsky, Kr—Kordonsky, Ir—Irensky

Fig. 2.3 Snow accumulation in sinkholes: **a** in the area of the Polazna Village on April 22, 2020; **b** in the Lukino Village 05/07/2020 (*photo by* N. Maksimovich)

At the same time, the considerable duration of the period with negative temperatures, in which there is practically no infiltration, as well as rather large snow volume and intense snow melting in spring, creates a significant unevenness in the activity of karst processes throughout the year.

Surface waters. The Perm Region ranks first in the Urals in terms of the availability of water resources. A well-developed hydrological network is represented by all the main types of inland water bodies: rivers, reservoirs, ponds, lakes, and swamps. The average long-term surface runoff resources are more than 57 km^3/year (1.3% of the river runoff volume in Russia). The density of the river network varies in a wide range: from 0.8 to 1.0 km/km^2 in the north of the region to 0.2–0.3 km/km^2 in the southern regions, averaging 0.5–0.6 km/km^2. The nature of the feeding of surface water bodies, mainly of the snow type, with a pronounced spring flood, summer-autumn rainfall floods, and a stable winter low-water period. Ground feeding forms, on average, 25–35% of the annual runoff (*Surface water* ... 1967).

Basic information about the rivers flowing through the regions of sulfate karst development is given in Table 2.1, all of them refer to the basin of the Kama Reservoir. Its filling began in 1954, and in the spring of 1956 the water level was brought up to the design mark of 108.0 m. Since 1961, the mark of the normal retaining level (NRL) has increased for economic reasons to 108.5 m. The surface area at the NRL is 1915 km^2, the volume of the water mass is 12.2 km^3, and the maximum depth is 32 m. During the winter operational drawdown, the level of the Kama Reservoir decreases by 7–8 m.

The Kama Reservoir is characterized by a complex morphology: a significant indented coastline, a sharp

Table 2.1 Rivers with a length of at least 200 km in areas of sulfate karst development (Scientific and applied ... 2015)

River	Confluence	River length, km	Basin area, km^2	River network density, km/km^2
Pilva	Kama	214	2020	0.65
Vishera	Kama Reservoir (Vishersky Bay)	415	31 200	0.64
Kolva	Vishera	460	13 500	0.72
Yazva	Vishera	162	5900	0.66
Glukhaya Vilva	Yazva	234	1740	0.81
Yayva	Kama Reservoir	304	6250	0.63
Kosva	Kama Reservoir (Kosvinsky Bay)	283	6300	0.58
Chusovaya	Kama Reservoir	592	23 000	0.69
Sylva	Kama Reservoir	493	19 700	0.55
Barda	Sylva	209	1970	0.79
Iren	Sylva	214	6110	0.45

alternation of widening and narrowing, and abrupt changes in depths, which causes certain differences in the hydrological regime of its individual sections. The length of the reservoir along the navigable route is 300 km.

Numerous outcrops of gypsum and anhydrite are noted on the shores of the Kamsky reach at the Dobryanka Town. Their dissolution affects the chemical composition of the water. Some increase in water salinity is noted in a narrow coastal strip near the Polazna Village, which is caused by the dissolution of sulfate rocks. In the dam part of the reservoir, the salinity is always higher than in the Khokhlovka, Polazna, or Dobryanka areas (70–450 mg/dm^3), which is caused by the mixing of the Kama waters with the more saline Chusov waters (Dvinskikh and Kitaev 2008).

The waters of the Chusovsky reach have a salinity of 100–400 mg/dm^3 before the confluence with the Sylvensky reach, which increases, reaching 500–600 mg/dm^3, below the confluence. The waters of the Sylvinsky reach are more mineralized (up to 1000 mg/dm^3).

The influence of karst is more pronounced on small rivers, reflected in the chemical composition and flow rates of rivers. They are characterized by the absorption of part or all of the runoff with reappearance downstream. Dry lands are typical forms in the regions of karst development.

The chemical composition of river waters in gypsum zones is characterized by a significant increase in calcium sulfates.

Mineralization of karst rivers exceeds the average value of river water salinity, which, according to O. A. Alekin, for the Caspian Sea basin is 267 mg/dm^3 (Gorbunova 1965).

The territory under consideration contains a large number of lakes, often of karst origin. The largest lakes of the Perm Region are associated with erosion-karst lowlands: Chusovskoe (19.4 km^2), Nyukhti (6.3 km^2), and others. The deepest lake in the region, Rogalek (61 m), is located in the Durnyatskaya Depression, a flowing karst depression.

Karst depressions are associated with vast wetlands on the watersheds of the Kama and Chusovaya, Sylva and Iren, Kolva and Vishera Rivers.

Soils. The physical and geographical conditions of the territories of the development of sulfate karst determined the predominance of the podzolic process of soil formation. Sod and bog processes are much less common (Specially protected… 2002). The variety of soil types is due to the heterogeneity of the parent rocks, the difference in elevation marks, and the originality of vegetation and climatic conditions.

In the northern regions, podzolic, partly bog-podzolic types of soils have formed on relatively poor-base mantle loams. Sod-podzolic soils were formed in the central and southern areas on eluvial-diluvial clays and loams. In the southern regions, soddy-calcareous soils were formed on calcareous parent rocks. A special place is occupied by soils of the Kungur island forest-steppe-podzolized and leached chernozems and forest-steppe dark gray, gray and light gray podzolized soils. Throughout the study area, alluvial sediments are developed in the river floodplains, and on the slopes and bottoms of ravines, gullies, in the floodplains of small rivers, on the steep slopes of ridges and river valleys.

The study area belongs to the zone characterized by the presence of seasonal permafrost, which occurs annually during the cold season. During the period of soil freezing, the supply of groundwater due to the infiltration of atmospheric precipitation stops, and the runoff is determined by the reserves of groundwater.

The features of the soil layer have an indirect effect on karst processes, mainly through the mechanisms of infiltration and evaporation. Low-mineralized atmospheric moisture, passing through the soil layer, dissolves the salts contained in the soil and carries them into the aquifer, changing the chemical composition of groundwater. At a lower thickness of the forest undergrowth, the infiltration of atmospheric precipitation into the soil occurs faster, and the amount of evaporation from the soil surface also increases. An increase in the thickness of the arable layer slows down the process of infiltration of atmospheric precipitation and melt water and increases the amount of evaporation from the soil surface.

The **vegetation** of the described territory is diverse, the predominant type is forests. Almost all regions of sulfate karst development are located in the zone of the middle and southern taiga. The main tree species in the middle taiga are spruce and fir. In the northern regions, spruce-fir forests are widespread in large continuous areas; mosses predominate in the above-ground cover. Such forests retain moisture, infiltration and evaporation are slowed down.

Spruce and fir form the basis of the southern taiga forests too. Southward, the share of deciduous species gradually increases, the undergrowth, shrub layer, and herbaceous cover change. Small-leaved birch-aspen forests, both of primary and secondary origin, occupy a fairly large proportion of tree plantations; broad-leaved trees (linden, elm, and maple) appear. To the south of the latitude of the Berezniki Town, the moss cover is displaced by herbaceous vegetation. Southern taiga forests also restrain moisture infiltration and evaporation, but to a lesser extent than middle taiga forests (Ovesnov 1997).

A special position is occupied by the region of the insular Kungur forest-steppe (between the Sylva and Iren Rivers; Fig. 2.4). The area is highly cultivated, small islands of forests have been preserved mainly along the hills, steppe areas are found both on the watersheds and on the slopes. High bedrock river banks are characterized by outcrops of gypsum, alternating with areas covered with meadow and forest vegetation (Specially protected … 2002).

Fig. 2.4 Valley of the Sylva River near the Kashirino Village (*photo by* N. Maksimovich)

At present, as a result of felling, significant areas are covered with secondary birch and mixed forests.

In the sites of development of soluble sulfate rocks with excessive moisture and the absence of woody and herbaceous vegetation, which inhibits infiltration and evaporation, favorable conditions are created for the development of karst processes.

Landscapes. Landscape transitions in the Perm Region are most pronounced when moving from north to south. Within the study area, three landscape units are distinguished, primarily due to the spread of vegetation. The northern part of the territory (approximately up to the latitude of the Solikamsk Town) is represented by mid-taiga landscapes; to the south, there are typical southern taiga (approximately to the latitude of the Kungur Town), which are gradually replaced by subtaiga (broad-leaved-coniferous). A unique phenomenon for the described territory, and the Perm Region as a whole, is the presence of landscapes of the forest-steppe type, which received the name "Kungur forest-steppe" (Tartakovsky 2012; Specially protected … 2002).

The influence of karst processes is initially expressed in a change in the microrelief of the landscape. As they develop, karst landscapes can become more and more isolated. The conditions of karst formation are of great importance: karst rocks come to the surface (bare karst), or are covered with a soil-sod layer (choked karst), or overlain by non-karst rocks (buried karst). The greatest impact on landscapes is usually associated with the bare karst, the smallest—with the buried one. To one degree or another, karst affects all elements of the landscape and is considered as a system of complex processes:

- geochemical and biochemical processes of migration of chemical elements in the rock-water system;
- the geological process of rock leaching, changes in its composition, structure, texture, the formation of new minerals, sediments, sedimentary rocks, and minerals;
- hydrogeological process of formation of reservoirs and chemical composition of groundwater;
- geomorphological process of formation of surface and underground landforms;

- physical and geographical process of formation of a special type of landscape;
- an engineering-geological process leading to a change in the strength properties and stability of territories, an increase in the degree of aggressiveness of groundwater in relation to the foundations of structures (Gorbunova et al. 1992; Gorbunova 1985).

In sites of outcrops of karst rocks, various karst landforms have developed, which largely determine the appearance of landscapes.

2.2 Tectonics and Geological Conditions

The territory of the Perm Region (area 160.6 thousand km^2) is located within three large geostructures: the eastern edge of the East European platform, the Uralian trough, and the folded Urals.

The presence in the eastern part of the region of the mountain range—the Urals—left an imprint on the peculiarities of the formation of the East European Platform adjacent to it from the west. Its eastern edge was involved in a mobile fold belt, and the crystalline basement sank here to a depth of 6–8 km (usually 2–3 km). The resulting Cis-Ural foredeep was gradually filled with alternating layers of marine sedimentary rocks and the products of destruction of the Ural Mountains.

An equally important aspect of the participation of the Urals in the formation of the eastern margin of the East European Platform is the Uralic origin of the loose material that composes the sedimentary cover strata. Numerous rivers, flowing down over many tens and hundreds of millions of years from the ancient Ural Mountains, transported with their waters a huge amount of sandy-clay material for hundreds of kilometers deep into the plain adjacent to the west.

In the platform part, under the Quaternary deposits, there are mainly Upper Permian deposits, represented by sandstones, marls, limestones, and conglomerates. Lower Permian carbonate and sulfate sediments come to the surface within the Kosvinsko-Chusovskaya Saddle, Bymsko-Kungurskaya Depression, Bashkir Arch, and Ksenofontovo-Kolvinskaya Saddle (north of the region).

In the Solikamsk Depression of the Cis-Ural Trough, a thick halogen stratum (rock salt, sylvinite, carnallite, gypsum, anhydrite, dolomite) is developed, overlain by Lower Permian (Ufimian stage) marls, platy limestones, and clays. In the southern and eastern parts of the trough, there are Lower Permian (Kungurian) sandstones, marls with lenses of limestones and conglomerates, mudstones, siltstones with lenses of gypsum, anhydrite, and rock salt.

The folded Urals are composed of dislocated terrigenous and carbonate rocks from the oldest Proterozoic to Lower Permian, complicated by meridionally elongated thrusts and other faults. The variety of geological environments, as well as the development of karst processes, determines the complexity of hydrogeological and engineering-geological conditions that affect the course of technogenic and natural-technogenic processes.

On a significant part of the Perm Region (47.5 thousand km^2), karst rocks come to the surface or lie shallow from it: limestone, dolomitized limestone, dolomite, gypsum, anhydrite, and salt. Modern and ancient karst is widespread in them.

The Perm Region is characterized by a meridional zonal structure. From west to east, the eastern part of the East European platform is replaced by the Cis-Ural trough and then by the geosynclinal-folded region of the Hercynian folded Urals. The eastern edge of the platform, where karst is developed, is complicated by large structures: the Timan Ridge, the Perm Arch, the Bym-Kungur Depression, and the northern end of the Bashkir Arch, within which small structures have been identified. The Timan Ridge and the Perm Arch continue in the Cis-Ural Trough with transverse structures: the Ksenofontovo-Kolvinskaya and Kosvinsko-Chusovskaya Saddles, which divide the trough into the Upper Pechora Depression, the Solikamskaya and Yuryuzan-Sylvinskaya Depressions.

The Hercynian folded Ural includes the West Ural outer folding zone and the Central Ural uplift.

Geological structure. The Paleozoic and Proterozoic deposits lie on the basement within the Perm Region. In the Upper Proterozoic (Riphean, Vendian), along with sandstones, mudstones, conglomerates, dolomites, limestones, marls, and volcanic rocks are present (Fig. 2.5).

The Silurian system (lower section, Kolchim Formation) occurs in the western folds of the Urals. It is composed of dolomites, dolomitized limestones, at its base is a layer of terrigenous rocks. The upper section of the Silurian (Yazvinskaya Formation) is developed in the extreme western folds of the Urals. It is represented by limestones, marls, mudstones, siltstones, and sandstones.

The Devonian system includes the middle and upper sections. The Eifelian and Givetian stages of the Middle Devonian contain both terrigenous and carbonate rocks. The Frasnian and Famennian stages of the Upper Devonian, with the exception of the Lower Frasnian substage, are composed mainly of carbonate rocks.

The coal system comes to the surface in the Ksenofontovo-Kolvinskaya Saddle near the Ksenofontovo Village and in the folded zone. Basically, it consists of carbonate rocks with a thickness of 250–1200 m, with terrigenous ones playing a subordinate role.

The Permian system is distributed mainly in the platform part and trough. The Asselian, Sakmarian, and Artinskian stages of the Lower Permian are dominated by limestones

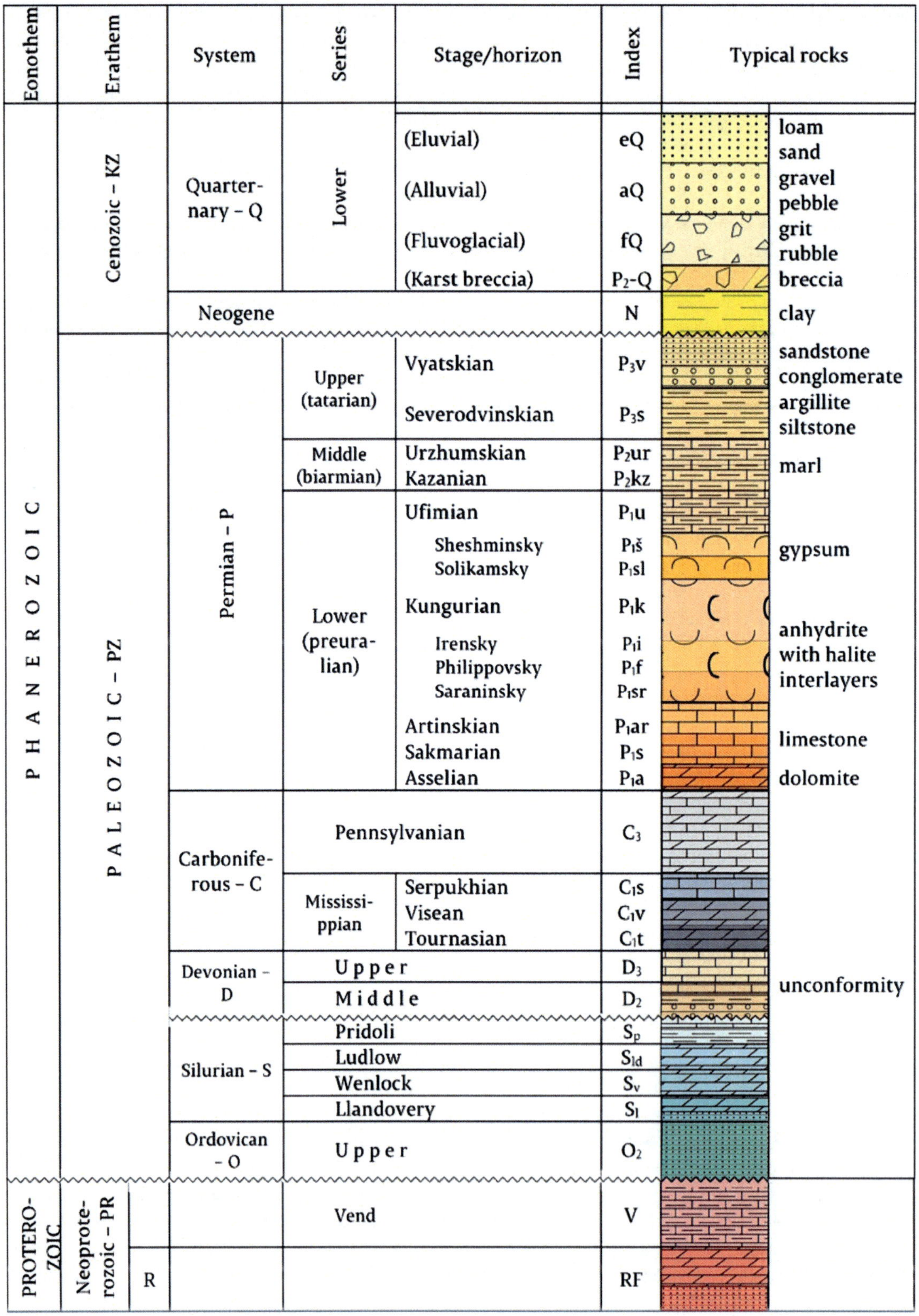

Fig. 2.5 The position of karst and rocks in contact with it in the stratigraphic section (according to Gorbunova et al. 1992)

and dolomites. The Artinsky reef massifs are developed on the border with the trough. Anhydrites are present in the Artinskian stage. In the trough, the upper part of the stage is represented by the Divya Formation of clays, siltstones, marls, and clayey limestones, which is replaced to the east by the more coarse-detrital Urma Formation.

The Kungurian stage is developed in the eastern part of the platform and trough. The Philippovsky and Irensky Horizons are distinguished in it.

The lower Solikamsky Horizon of the Ufimian stage is composed of limestones, marls, dolomites with interlayers of anhydrite, gypsum, and rock salt. The upper part contains sandstones, siltstones, and clays. Above is the Sheshminsky Horizon (sandstones, mudstones, siltstones with gypsum veins, interlayers of limestone, and marls).

The Quaternary system consists of eluvial, diluvial, alluvial, lacustrine, boggy, fluvioglacial and glacial clays, loams, sands, pebbles, and peat deposits. The glacier of the Dnieper glaciation covered only the northern part of the region.

Thus, carbonate rocks appear in the sections of the Upper Proterozoic. They reach the greatest development in the Carboniferous system. Sulfate manifestations are noted in the Artinskian stage, but prevail in the Kungurian. Salts are also typical for the Kungurian stage (Gorbunova et al. 1992).

2.3 Hydrogeological Conditions

In terms of hydrogeological zoning, the western territory of the Perm Region belongs to the East Russian and Ural complex basins of formation waters; the eastern part (mountainous Urals)—to the Bolsheuralsky complex basin of crustal-block (bed-block) and formation waters (Mineral resources … 2006).

The main distribution of karst rocks is confined to the Predural and Bolsheural Basins. Sulfate karst is associated with the aquifer of Permian sediments.

Aquiferous sedimentary complex of the Sheshminsky horizon. The Sheshminskaya formation is represented by sandstones, siltstones, and clays. Its characteristic feature is plaster cast; gypsum is found in cement and in the form of veins or individual crystals in sandstones, clays, and siltstones.

The chemical composition of the underground waters of the Sheshminsky complex in the zone of active water exchange is hydrocarbonate-sulfate–calcium-magnesium with a mineralization of 0.1–0.5 g/dm^3, which increases with depth, which leads to a transition to a sulfate composition. Below the erosional incision, the gypsum rocks are not washed out, and the waters in them are pressurized with high salinity, sulfate–calcium, and chloride-sodium composition.

With a technogenic change in the filtration regime, for example, at the base of the Kama HPP, signs of dissolution of the gypsum rocks of the Sheshminsky horizon were observed.

On the flanks of the Yara uplift, there is a zone with an increase in the amount of sulfates. The inflow of saline sulfate and chloride waters from the underlying horizons along the fractured zones of the swells also occurs in the Verkhnekamskoye potassium-magnesium salt deposit.

The aquifer of the deposits of the Solikamsky formation is composed of a stratum of marls, limestones, sandstones with interlayers of clays, and siltstones; there are interlayers of gypsum and anhydrite. The Solikams deposits are flooded very unevenly. The presence of waterless areas was established in areas where the Solikams deposits are not very thick and are underlain by a strongly karsted gypsum-anhydrite stratum of the Irensky formation, which completely drains the overlying rocks.

The uneven fracturing of the rocks provided the unequal water cut of the complex. The flow rates of the springs range from 0.1 to 50 l/s, less often up to 270 l/s.

The chemical composition of waters is diverse: hydrocarbonate, sulfate, and chloride. Sulfate waters with salinity from 0.4 to 3.0 g/dm^3 are more typical for the lower, less washed clay-marl strata of the Cis-Ural Trough. To the south of the Bogorodskoye Village, there are mainly sulfate or sulfate-hydrocarbonate waters with salinity from 1.0 to 3.0 g/dm^3, which is associated with an increase in the role of chemogenic sediments in the rock mass.

Aquiferous sedimentary complex of the Irensky suite. The Irenskaya formation of the Kungurian stage is represented by alternating gypsum-anhydrite and carbonate units and comes to the surface in the wings of the Perm-Bashkir Arch. Irensky gypsums occupy rather wide areas and are heavily karst. Hypsoanhydrite and carbonate rocks of the Irenskaya suite are water-bearing only at the point of their emergence to the surface or when they occur very close to the surface. With the immersion of the Irensky rocks under the younger sediments, karst processes damp in them and the water content sharply decreases.

In the sites where the Irenskaya suite emerges on the surface, disconnected karst streams with different piezometric levels and sporadic water content prevail. On the western wing of the Perm-Bashkir Arch in the Lunezhskaya Aquifer, karst waters have a single mirror and represent a zone of complete saturation. On the watershed of the Babka and Sylva Rivers and on the right bank of the Sylva River, the Lunezh gypsum is not water-bearing.

As a result of intensive karst destruction of the lower part of the Irenskaya suite, peculiar deposits are formed—the Olkhovskaya karst breccia, it is waterless, since the waters completely seep through it into the underlying rocks.

The karst waters of the Irenskaya suite are characterized by a pronounced sulfate–calcium composition, increased mineralization (2.5–3.0 g/dm^3) and hardness (up to 30 meq/dm^3), which makes them unsuitable for domestic water supply.

Aquiferous sedimentary complex of the Kungurian stage (Popovskaya, Koshelevskaya, and Lekskaya formations). Within the complex, there are water-bearing sandstones, marls, limestones, gypsum lenses, conglomerates, and conglomerate-breccias. The aquicludes are represented by clays, sometimes sandstones, and also salt-bearing strata. The variety of lithological composition determined the presence of fractured-ground, fractured-stratal and fractured-karst waters, their complex relationship, and chemistry.

The water abundance of rocks is uneven. Typical flow rate of springs from sandstones and marls is 0.5–2 l/s, and of wells is 1.0–5.0 l/s.

The chemical composition of groundwater is very variegated. In the sites of the complex outcrops, hydrocarbonate and hydrocarbonate-sulfate waters with a salinity of 0.1–0.5 g/dm^3 are most widely developed. Sulfate waters with a salinity of 1–3 g/dm^3 are confined to the lenses of gypsum and conglomerate. Chloride waters mineralized up to 300 g/dm^3 are associated with rock salt leaching.

Aquiferous sedimentary complex of the Artinskian-Kungurian stages is composed of limestones, dolomitized limestones, and dolomites, and interlayers of anhydrites appear in the sites of outcrops.

Within the Perm-Bashkir Arch, the Artinskin-Kungurian Horizon forms a complex basin of fissure-karst waters with a bilateral runoff. Groundwater, filling cracks, and voids form a single aquifer.

The water abundance of rocks is different: in the places where the horizon reaches the surface, the background flow rates of springs are 0.5–1 l/s, of wells are 0.5–2 l/s; in fractured water-rich zones, the flow rate of springs increases to 10–20 and even 500–700 l/s, and of wells up to 10–20 l/s.

In the sites where the horizon reaches the surface, the mineralization of karst waters reaches 1 g/dm^3, the composition is hydrocarbonate-calcium. With an increase in the depth of water and with the immersion of aquifers under the Irensky gypsum, mineralization increases to 3 g/dm^3 and more the composition changes to sulfate and sulfate–chloride, and hydrogen sulfide appears.

The waters of the Artinskian-Kungurian horizon at the point where it comes to the surface are widely used for drinking and technical water supply. With immersion under the Irensky gypsum-anhydrite stratum of the horizon, they acquire a high mineralization, sulfate, more often chloride composition. Waters with a high content of hydrogen sulfide are used for balneological purposes (Bol. Klyuchi Resort) (see Sect. 3.7).

Aquiferous sedimentary complex of the Asselian-Artinskian stages is composed of sandstones; mudstones; and interlayers and lenses of conglomerates, limestones, and marls, with no plastering at all. The water abundance of the rocks is not the same. Mineralization of groundwater is low (0.15–0.4 g/dm^3), hydrocarbonate-calcium in composition, of moderate hardness (Hydrogeology of the USSR … 1972).

2.4 Karst Phenomena and Forms

On the territory of the Perm Region, there are all types of karst, distinguished by lithological composition: carbonate, salt, sulfate, and their combinations (Gorbunova et al. 1992; Kataev et al. 2013). There are also complex types of karst in the case of genetic heterogeneity (bedrock, eluvium, alluvium, etc.) of the cover sediments (Fig. 2.6).

The results of the manifestation of the karst process in various geological-hydrogeological and physical-geographical conditions are karst forms, which are divided into surface, underground, and transitional.

Data on karst forms, surface, transitional, and underground, as well as on karst sources confined to areas of development of sulfate karst, are presented in Table 2.2.

The greatest danger to life is posed by sinkholes—karst funnels with a known date of formation. They arise when the roof of a cavity formed in loose cover deposits as a result of soil removal along cracks, channels into karst rocks; they are often found on terraces, where cover deposits are represented by loams, clays, sands, and pebbles (Gorbunova 1990; Gorbunova et al. 1990). In the Perm Region, 70% of the phenomena associated with deformations of the earth's surface and including slow subsidence and catastrophic sinkholes are associated with sulfate karst.

Dynamics of karst sinkholes on the territory of regions of predominantly sulfate karst of the Perm Region. In the regions of distribution of sulfate deposits on the territory of the Perm Region, the most common forms of hazardous surface manifestations of karst processes are sinkholes. There are 11 regions with sulfate deposits (Gorbunova et al. 1992). The Kungur laboratory of the Mining Institute of the Ural Branch of the Russian Academy of Sciences recorded gaps within the most populated territories of five regions: Polaznensky, Irensky, Kishertsky, Nizhnesylvinsky, and Lysvensky (Fig. 2.7).

The number of sinkholes by region is presented in Table 2.3. In the Kishertsky region, over 100 years, there were 13 sinkholes per 1 km^2, 5 in the Irensky and Nizhnesylvinsky regions each, 8 in the Lysvensky region, and 3 in the Polaznensky region.

The largest number of sinkholes was recorded in the period 1960–1980 (Fig. 2.8). Probably, the increase in the number of sinkholes in the Polaznensky and

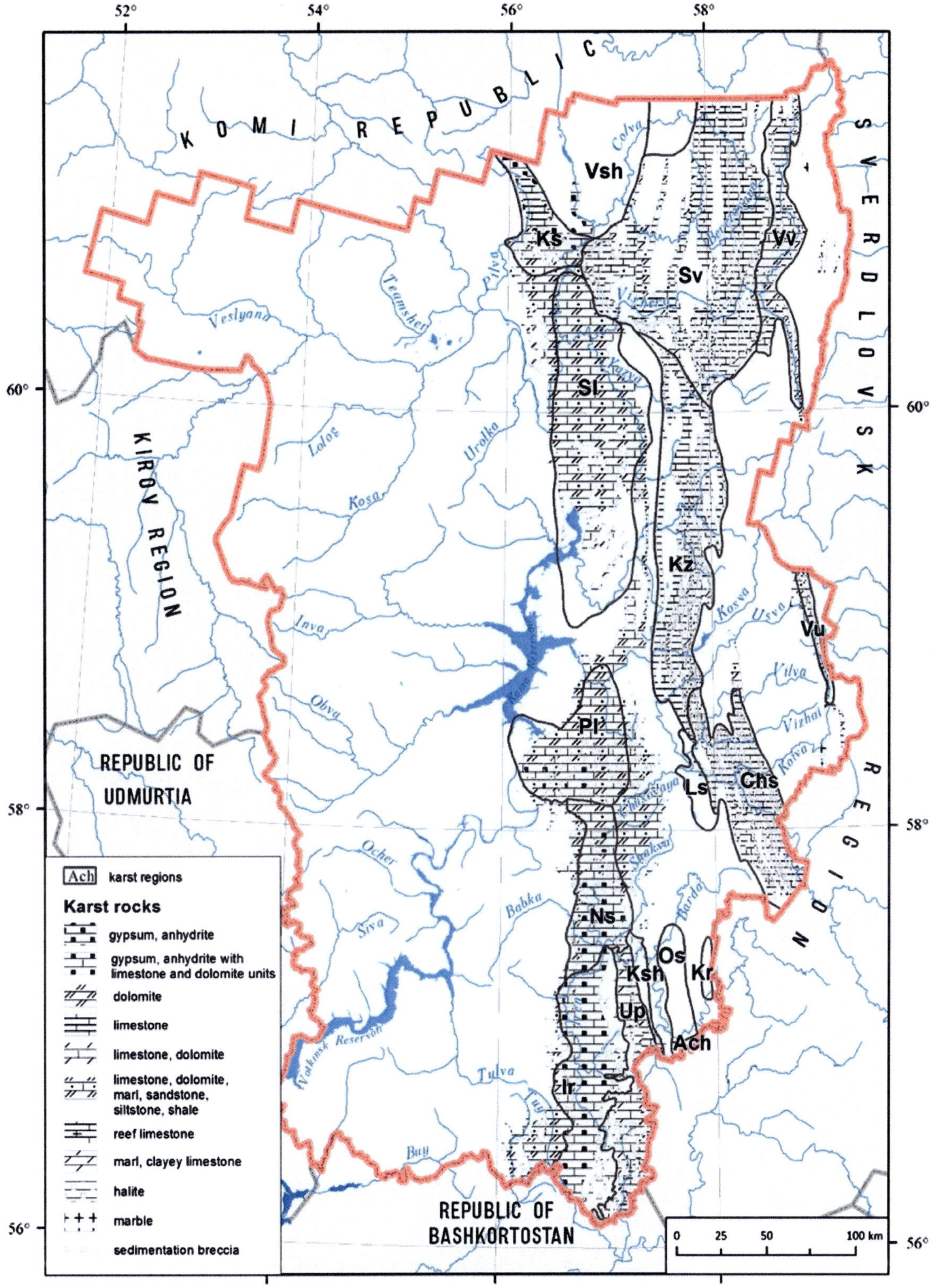

Fig. 2.6 Map of karst rocks and karst of the Perm Region (Tartakovsky 2012; Gorbunova et al. 1992). *Note* In some regions, karst rocks do not have continuous distribution. Karst regions: Ks—Ksenofontovsky, Vsh—Visherksky, Sv—Srednevishersky, Vv—Verkhnevishersky, Sl—Solikamsky, Kz—Kizelovsky, Vu—Verkhneusvinsky, Pl—Polaznensky, Ls—Lysvensky, Chs—Chusovskaya, Ns—Nizhnesylvinsky plateau, Ksh—Kishertsky, Os—Osintsevsky, Ach—Achitsky, Kr—Kordonsky, Ir—Irensky

Table 2.2 The main forms occurring in the areas of sulfate karst development in the Perm Territory (Atlas of electronic … 2000 with additions by the authors)

Karst regions	Karst forms															
	Surface									Transitional, underground						
	Depressions	Subsidence troughs	Kettles	Sinkholes	Dolines	Lakes	Gulches	Ravines	Karrens	Caves	Wells	Ponors	Cavities	Channels	Karst joints	Springs
Ksenofontovsky	+		+	+	+		+	+		?		+		+	+	+
Visherksky	+		+	+		+	+	+	+			+		+	+	+
Solikamsky	+	+	+	+	+	+	+	+				+	+	+	+	+
Polaznensky	+		+	+	+	+	+	+	+	+		+	+	+	+	+
Lysvensky			+	+	+	+	+	+	+	+		+	+	+		+
Nizhnesylvinsky	+		+	+	+	+	+	+	+	+	+	+	+	+	+	+
Kishertsky	+		+	+	+	+	+	+	+	+		+	+	+	+	+
Osintsevsky	+		+	+	+	+	+	+						+	+	+
Kordonsky				+	+		+	+					+	+	+	+
Achitsky				+			+	+						+		+
Irensky	+		+	+	+	+	+	+	+	+	+	+	+	+	+	+

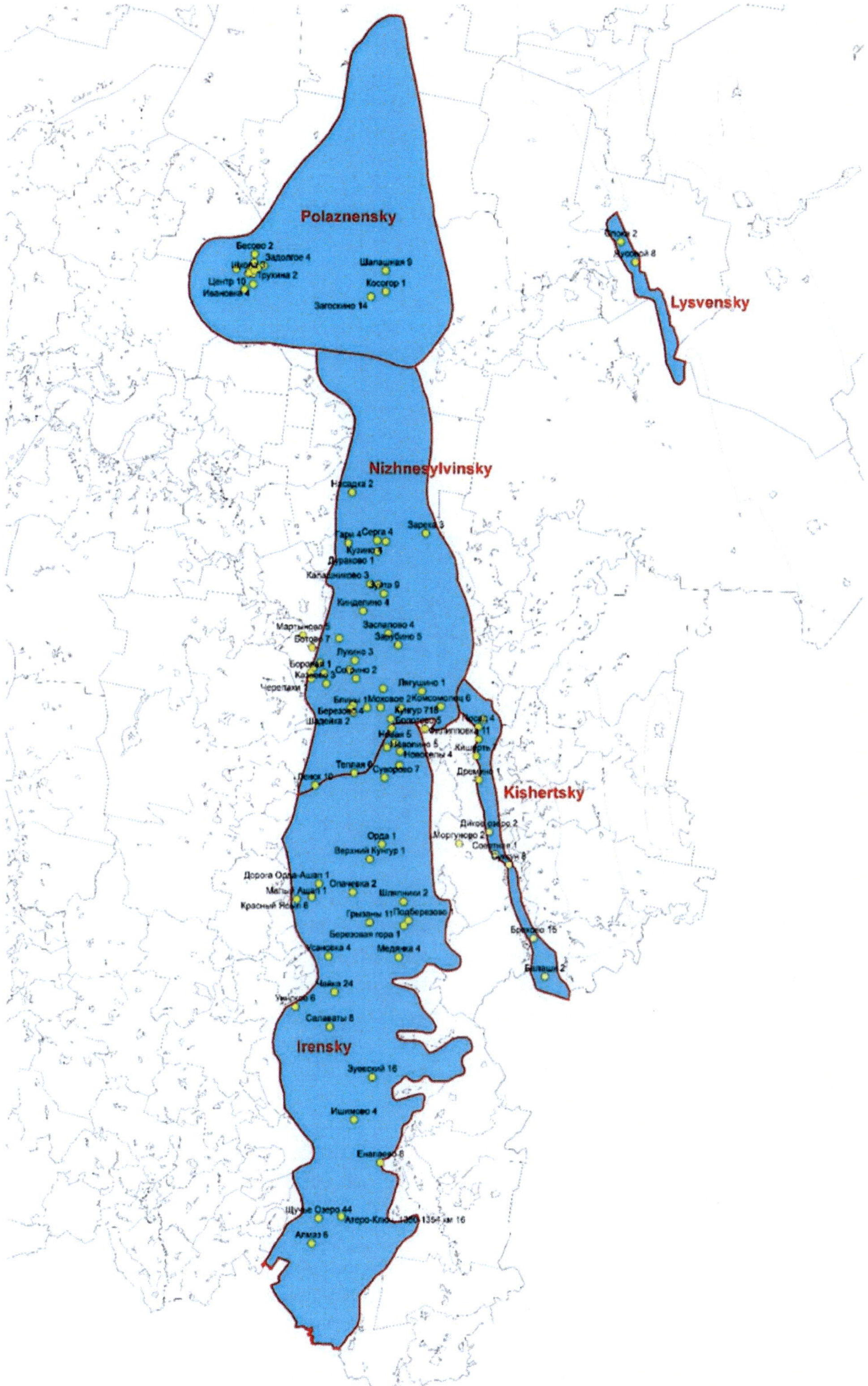

Fig. 2.7 Areal distribution of sinkholes over the past 100 years on the territory of some regions of the sulfate karst of the Perm Region

Table 2.3 Sinkholes in the regions of sulfate karst

Karst region	Region area, km^2	Number of sinkholes, over 100 years	Number of sinkholes, per 1 km^2
Polaznensky	3102	79	3.33
Irensky	3815	197	5.83
Nizhnesylvinsky	2191	128	5.88
Kishertsky	334	49	13.8
Lysvensky	505	10	8.6
Total	9947	463	–

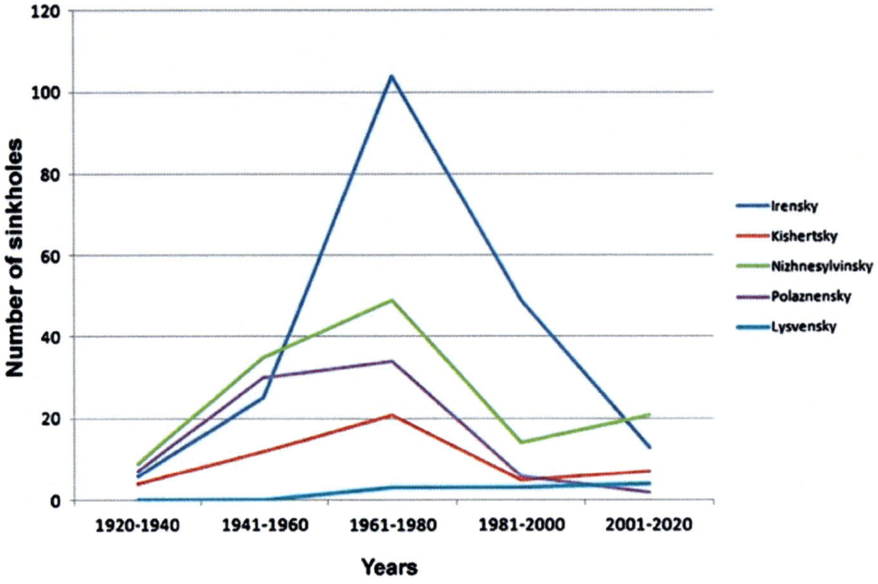

Fig. 2.8 The number of sinkholes in some regions of the sulfate karst of the Perm Region

Nizhnesylvinsky regions during this period was associated with a rise in the level of surface and groundwater after the construction of the Kama HPP in 1954, and in Irensky and Kishertsky—with areal plowing of agricultural land.

Karst sinkholes are the most dynamically developing forms of surface manifestation of the process. It is their potential catastrophic nature that basically determines the concept of karst hazard. The formation of such sinkholes creates threats to life in the urbanized area and the operational safety of engineering facilities. Karst hazard is determined not only by the intensity of formation of sinkholes, but also by their size, primarily by their diameter.

The two sinkholes largest in diameter occurred in 1949 in the Ust-Khert Village: the first with a diameter of 120 m and a depth of 27 m and the second with dimensions of 60 by 40 m and a depth of 15 m. The deepest sinkhole was recorded in the Brekhovo Village (Kishertsky karst region) in 1953, its initial depth (at the time of the survey) was 40 m.

The total area of sulfate karst distribution regions, where sinkholes are most frequent, is about 10 thousand km^2. Monitoring observation data within the settlements made it possible to identify sites of the first hazard category. The area of the identified sites of the first hazard category is 18 km^2, which is 7% of the area of urbanized territories (261.74 km^2) in the regions of carbonate-sulfate karst distribution in the Perm Region. Information about some sinkholes is given in the description of the regions of development of sulfate karst (see Chap. 3).

2.5 Sulfate Karst in Conditions of Technogenic Impact

In areas with a significant technogenic load, the conditions and factors of karst formation change significantly due to irreversible transformations of the relief and rocks, pollution of surface and ground waters, the atmosphere, degradation of vegetation, etc. (Gorbunova et al. 1990, 1994; Gorbunova and Maksimovich 1991).

In 1956, K. A. Gorbunova made an attempt at engineering-geological zoning of the karst territory of the Perm Region (Gorbunova 1956), according to which four categories of territories were distinguished, differing in varying degrees of stability for engineering structures:

1st category (I)—territories that are unfavorable in engineering and geological terms (regions of predominantly sulfate karst);

2nd category (II)—territories that are relatively favorable in engineering and geological terms (regions of karst development in carbonate rocks);

3rd category (III)—territories with small sites, unfavorable in engineering and geological terms (Cis-Ural trough);

4th category (IV)—favorable territories, no karst.

K. A. Gorbunova (1990) also estimated the karst-hazardousness and the possibility of emergencies in karst territories. On this basis, five groups of regions are outlined, of which the regions of development of sulfate karst are included in the first three.

1. Regions of intense sulfate (gypsum-anhydrite) platform-type karst (Ksenofontovsky, Polaznensky, Nizhnesylvensky, Irensky, Kishertsky, Lysvensky [separated by the authors; according to K. A. Gorbunova—karst sites]). Construction conditions are determined by the presence of sites from I (very unstable) to IV (with slightly reduced stability) categories. The activation of karst is noted in connection with the technogenic impact. The occurrence of sinkholes can create emergency situations. Water supply conditions are unfavorable.
2. Regions of covered salt karst of troughs and depressions (Visherksky, Solikamsky), where the activation of karst is noted in old salt fields, during the exploitation of subsalt deposits, drilling wells, salt extraction, and intensive water intake. Negative processes: slow subsidence of the earth's surface (on a long-term scale), waterlogging, and salinization of groundwater in the zone of active water exchange. In places of intense technogenic load, emergency situations are possible when the roof of underground workings and karst cavities collapses.
3. Regions with local manifestations of covered sulfate (gypsum-anhydrite and salt) karst of small structures of depressions (Osintsevsky, Achitsky, Kordonsky). In places where gypsum is shallow, sinkholes may occur. Karst activation is likely in places where saline waters of lower horizons flow into the zone of active water exchange (self-flowing structural wells), as well as during engineering works accompanied by disturbance of cover sediments (Gorbunova 1990).

Sulfate karst affects all components of the natural landscape and human economic activity (Table 2.4; Gorbunova 1977).

The industry of the Perm Region is a powerful diversified economic complex. A significant part of the Kama region is associated with the extraction and processing of natural resources (Bereznikovsko-Solikamsky, Kizelovsko-Gubakhinsky, Lysvensko-Chusovsky industrial centers). In the north of the region, there are areas with logging and mining industries, in the south—areas with agricultural specialization. Key industries of the region: oil, chemical and petrochemical, ferrous and non-ferrous metallurgy, mechanical engineering, and timber industry (Fig. 2.9).

The greatest danger for engineering structures is represented by deformations of the earth's surface, which are very typical for regions with the development of sulfate karst.

The region has developed automobile and railroad networks, oil and gas pipelines (Fig. 2.9). Karst poses a great danger to the stability of railway structures, which is due to both the regional features of the development of this phenomenon (the meridional elongation of the regions of development of karst rocks and their rather large areal distribution), and the specifics of the design of railways (attraction to river valleys, soft conjugation of the track with the base, active dynamic impact on the base, linear continuity, etc.) (Yakovenko 1969).

The total length of railway lines in the Perm Region is 1574 km (Zaitsev et al. 2020), of which about 500 km (35%) pass through karst regions (Fig. 2.9). Engel's coefficient for the railway network reaches its maximum values (7.5–15.2 units) in the Polaznensky, Solikamsky, and Lysvensky Regions of the sulfate karst. Sinkholes in the railroad lane are especially dangerous and require study. In the last 40 years alone, there have been more than 20 such cases. The size of the sinkholes reached in diameter and depth from 5–9 to 20 m. The sinkhole in the Solikamsky karst region in the Berezniki Town in 2006, associated with the flooding of one of the potash mines, the dimensions of which were 337 by 202 m in plan, in bedrocks—305 by 170 m (2008), demanded the transfer of the railway.

More than 200 sinkholes were recorded within the boundaries of highways and their branches. Most of them were formed in ditches and in soil reserves and were caused by anthropogenic impact on the geological environment. At the same time, among the main factors activating karst and its manifestation on the surface, there is a decrease in the thickness of the overlying sediments up to the outcropping of karst rocks, a change in properties, in particular, with moisture, a transformation of the relief, leading to the appearance of foci of increased infiltration and inflation of surface, as a rule, very aggressive waters, and dynamic loads (Kostarev et al. 1990).

Pipeline transport has developed significantly in the Perm Region. A number of major oil and gas pipelines of national and international importance pass through the region. These are two routes of the Surgut-Polotsk oil pipeline, five main oil pipelines Urengoy-Center, the export gas pipeline Urengoy-Pomary-Uzhgorod (Nazarov and Sharygin 1999).

Table 2.4 Influence of sulfate karst on the landscape and human economic activity (Gorbunova 1977)

Terrain components	Process direction	Process consequence	Characteristic elements of the landscape	Impact on economic activity
Rocks: gypsum, anhydrite	Carrying out	Partial or complete leaching of rocks	Thin soils Local increase in the thickness of loose sediments Karst breccias Karst sinkholes, dolines, kettles, depressions Peculiar microclimate Spotted vegetation	Unfavorable geotechnical conditions Reduction of arable land Complications with the use of machinery in agriculture and forestry Deterioration of forest quality Recreational and other use
Surface and groundwater	Accumulation	Concentration of SO_4 and Ca ions in water; precipitation of $CaCO_3$ from water	Karst lakes	Use for breeding birds, fish farming, rarely for water supply
			Transfer of surface runoff to underground Small rivers of sulfate composition Sulfate karst springs Local watering Sulfated karst waters Sulfated groundwater	Unfavorable water supply conditions
			Hydrogen sulfide mud Hydrogen sulfide waters	Medicinal use

The length of trunk pipelines reaches 11.2 thousand km, of which 1.3 thousand km are oil pipelines and 9.9 thousand km are gas pipelines. About 22% of the oil produced in the country and 53% of natural gas pass through the territory of the region.

A special danger is posed by the technogenic activation of karst in conjunction with other dangerous geological processes, suffusion, erosion, flooding, landslides, etc., which led to a sharp intensification of sinkhole formation (Kostarev et al. 1990).

Karst regions are a place of increased risk for the construction of transport and communication network facilities.

There are a large number of accidents on main pipelines and railways associated with karst regions. Based on the analysis of accidents and damage to structures in karst regions, a conditional list of the severity of their consequences was compiled:

- catastrophic destruction (loss of general stability of the structure or damage to technological equipment), leading to the death of people or unacceptable contamination of the area with harmful chemical, radioactive, and other substances;
- partial destruction and damage leading to the temporary cessation of the normal operation of structures;
- damage leading to significant complications in the normal operation of structures;
- damage, under certain conditions leading to temporary difficulty in the normal operation of the structure; and
- damage that practically does not lead to difficulty in the normal operation of the structure (Tolmachev 1990).

Technogenic impact on the massifs of sulfate rocks significantly affects the intensification of the karst process. For example, the creation of the Kama reservoir caused the activation of the karst process, including at the base of the Kama HPP (Maksimovich 2006), and oil pollution of sulfate rocks led to the development of specific microbiological processes (see Sect. 3.4).

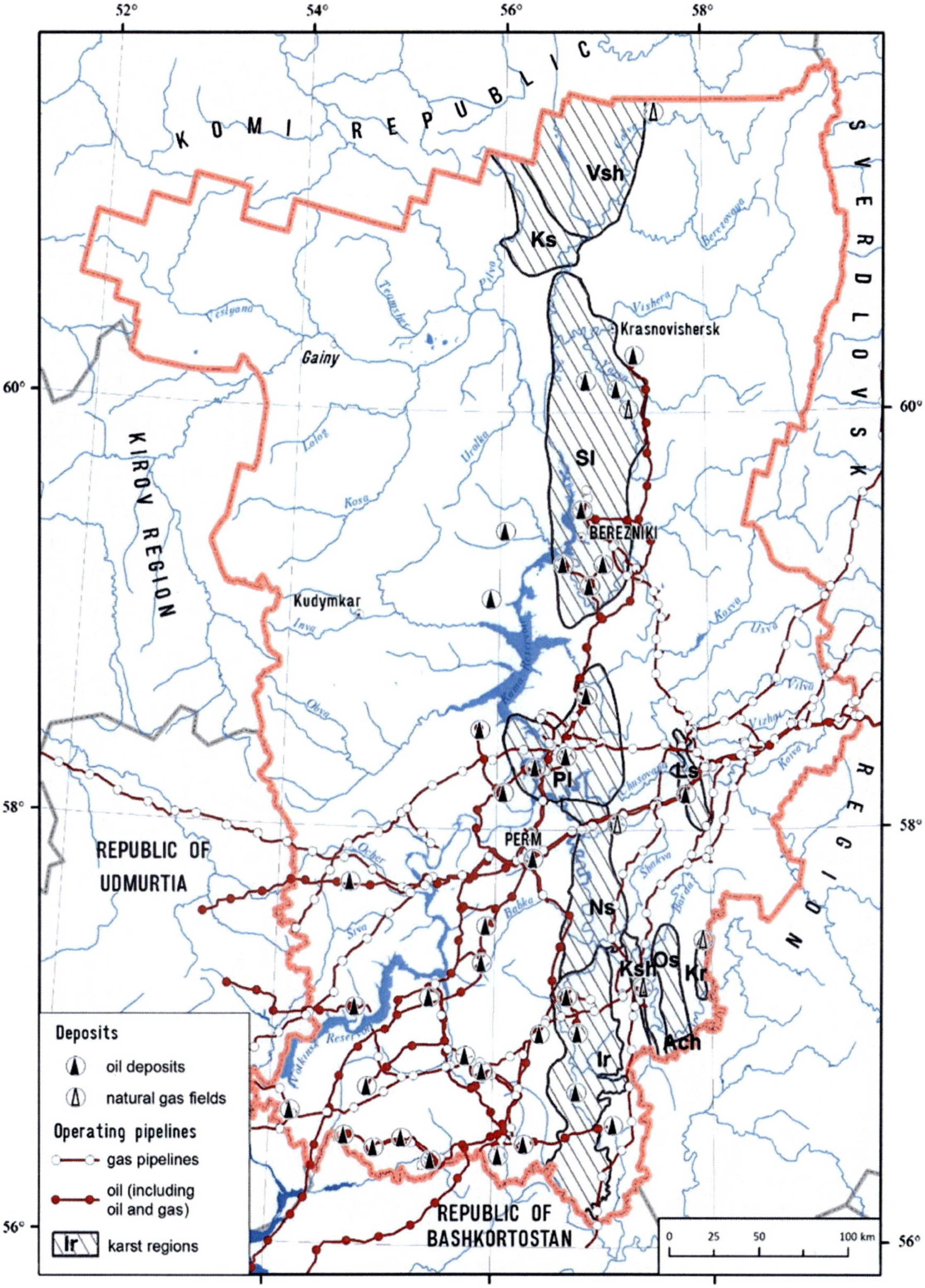

Fig. 2.9 Scheme of oil and gas fields and trunk pipelines in the sulfate karst regions of the Perm Region (Zaitsev et al. 2020; Gorbunova et al. 1992). Note: Regions of sulfate karst: Ks—Ksenofontovsky, Vsh—Visherksky, Sl—Solikamsky, Pl—Polaznensky, Ls—Lysvensky, Ns—Nizhnesylvinsky, Ksh—Kishertsky, Os—Osintsevsky, Ach—Achitsky, Kr—Kordonsky, Ir—Irensky

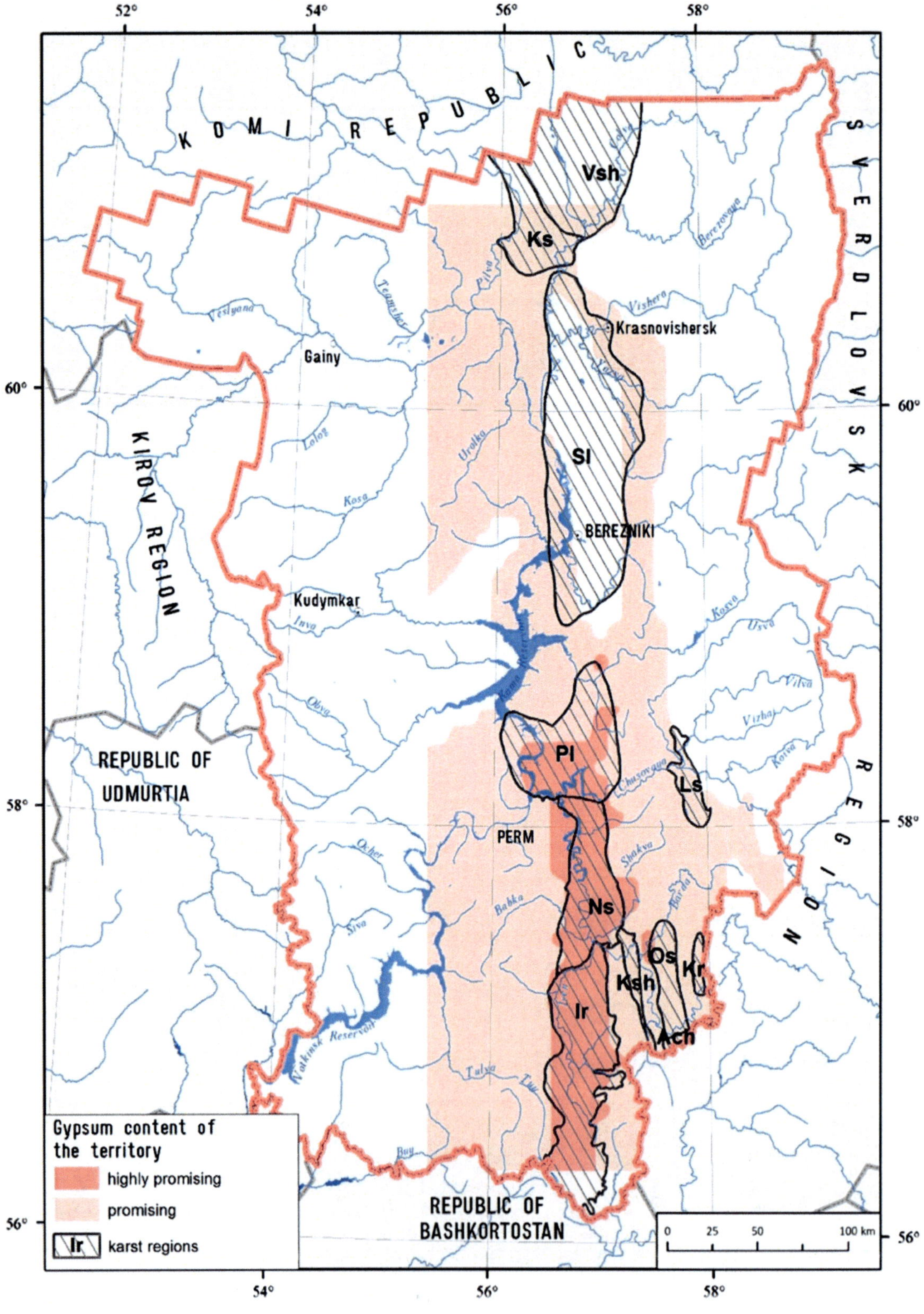

Fig. 2.10 Karst regions (Gorbunova et al. 1992) and the prospects of gypsum-bearing territories (Darovskikh and Kudryashov 2001). *Note* Karst regions: Ks—Ksenofontovsky, Vsh—Visherksky, Sl—Solikamsky, Pl—Polaznensky, Ls—Lysvensky, Ns—Nizhnesylvinsky, Ksh—Kishertsky, Os—Osintsevsky, Ach—Achitsky, Kr—Kordonsky, Ir—Irensky

The territories of gypsum deposits and promising areas in the Perm Region are confined to regions of development of sulfate karst (Fig. 2.10).

Karst has a negative impact on gypsum deposits, sometimes contributing to the destruction of a significant part of them, contaminating raw materials and complicating the mining conditions; therefore, when exploring them, it is extremely necessary to study the karst content of the territory (Zalkind and Nechaev 1959).

In the gypsum deposits of the Perm Region, there are five types of sulfate karst (Gorbunova 1977; Darovskikh and Kudryashov 2001):

- Karst develops in the surface part of the gypsum layer, covered only by loose sediments; the development of karst in this case depends on the nature of the relief of the site, the degree of permeability of the deposits overlying the gypsum layer, and on the amount of water entering its surface.
- A karst cavity develops along the top of the carbonate layer; characteristic features of this type of karst cavities are low thickness, significant dimensions in length and width; more often the cavities are not filled with anything.
- A karst cavity is formed under the base of the karst layer at the end of the sulfate layer; in this case, the cavity gets unlimited upward growth, since gypsum is washed out and dissolves easily; cavities are usually filled with loam with rubble and lumps.
- A karst cavity develops in the thickness of the sulfate layer along a thin carbonate layer; this type is a combination of the second and third types; and cavities are usually filled with rubble, lumps, and less often loam.

More detailed information on the technogenic impact on sulfate rocks and its consequences during economic activities are given in the description of the karst regions (see Chap. 3).

References

Atlas of electronic maps based on the ArcView GIS 3.2a software. (2000). "GeoMap".

Darovskikh, N.A., & Kudryashov, A.I. (2001). *Geology and prospecting for ornamental gypsum deposits*. Perm: GI UB RAS. (in Russian).

Dvinskikh, S.A., & Kitaev, A.B. (2008). *Hydrology of the Kama reservoirs*. Perm: PSU. (in Russian).

Engineering geology of the USSR. (1978). Russian platform. Vol. 1. M.: Publishing house of Moscow University. (in Russian).

Efimova, N.A., & Zubenyuk, L.I. (1966). Radiation and heat balance of the Urals. *Problems of physical geography of the Urals*. Moscow. (in Russian)

Gorbunova, K.A. (1956). *Karst of some regions of the Perm Region*: PhD thesis. Perm. (in Russian).

Gorbunova, K.A. (1965). Features of gypsum karst. Perm: Perm publishing house. http://nsi.psu.ru/cave/bibl/ogk.pdf. (in Russian).

Gorbunova, K.A. (1977). *Karst gypsum of the USSR*. Perm: Publishing house of Perm University. http://nsi.psu.ru/cave/bibl/kgu.pdf. (in Russian).

Gorbunova, K.A. (1985). *Karst studies. Issues of typology and morphology of karst*. Perm: Perm University. http://nsi.psu.ru/cave/bibl/kvtimk.pdf. (in Russian).

Gorbunova, K.A. (1990). Zoning of the territory of the Perm region by karst hazard. *Catastrophes and accidents in karst territories: abstracts of the meeting*, 27–28. (in Russian).

Gorbunova, K.A., Andreychuk, V.N., Kostarev, V.P., Maksimovich N. G. (1992). *Karst and caves of the Perm region*. Perm: Publishing house of Perm University. http://nsi.psu.ru/labs/gtp/stat/ng_0129.pdf. (in Russian).

Gorbunova, K.A., & Maksimovich, N.G. (1991). Technogenic impact on the karst territories of the Perm Region. *Geography and natural resources*, 3, 42–46. http://nsi.psu.ru/labs/gtp/stat/ng_0122.html. (in Russian).

Gorbunova, K.A., Maksimovich, N.G., Andreychuk, V.N. (1990). *Technogenic impact on the geological environment of the Perm region*. Sverdlovsk: Ural Branch of the Academy of Sciences of the USSR. http://nsi.psu.ru/labs/gtp/stat/109.pdf. (in Russian).

Gorbunova, K.A., Maksimovich, V.N., Kostarev, V.P. (1994). Technogenic activation of karst sinks in Perm region. *Proceeding 7 Int. Congress Ass. of Engineering Geology*. Portugal, Lisboa, 3, 1929–1931.

Hydrogeology of the USSR. (1972). Volume XIV. Ural. Ed. V.F. Preis. Ural Territorial Geological Administration. M.: "Nedra". (in Russian).

Kataev, V.N., Maksimovich, N.G., Meshcheryikova, O.Yu. (2013). Types of karst of the Perm Region. *Bulletin of the I. Kant Baltic Federal University*, 1, 56–66. http://nsi.psu.ru/labs/gtp/stat/2013/0398.pdf. (in Russian).

Kostarev, V.P., Dimukhametov, M.Sh., Papirovaya, V.T. (1990). Monitoring elements and emergencies at linear structures in the karst regions of the Perm Cis-Urals. *Catastrophes and accidents in karst territories: abstracts of the conference reports*, 31–32. (in Russian).

Maksimovich, N.G. (2006). *Safety of dams on soluble rocks (on the example of the Kama hydroelectric power station)*. Perm: OOO PS "Garmoniya". http://nsi.psu.ru/labs/gtp/stat/2006/0298.pdf. (in Russian).

Maksimovich, N.G., & Pyankov, S.V. (2018). *Kizel coal basin: ecological problems and solutions*. Perm: Perm State National Research University. http://nsi.psu.ru/labs/gtp/stat/2018/529.pdf. (in Russian).

Mineral resources of the Perm Region. (2006). Encyclopedia. A.I. Kudryashov (ed.). Perm: Publishing House "Knizhnaya ploshchad". (in Russian).

Nazarov, N.N., & Sharygin, M.D. (1999). *Geography. Perm region*. Perm: "Knizhnyy mir". (in Russian).

Ovesnov, S.A. (1997). *Synopsis of the flora of the Perm Region*. Perm. (in Russian).

Scientific and applied reference book on the climate of the USSR. (1990). Leningrad: Gidrometeoizdat. Issue. 9. (in Russian).

Scientific and applied reference book: Basic hydrological characteristics of the rivers of the Kama basin [Electronic resource] (2015). Georgievsky, V.Yu. (ed.). Electronic text and graphic data in PDF format (2 files: 25 MB). Livny: Publisher G.V. Mukhametov, 2015. (in Russian).

Specially protected natural territories of the Perm Region. (2017). Buzmakov, S.A. (ed.). Perm: Aster. (in Russian).

Specially protected territories of the Perm Region: register (2002). Ovesnov, S.A. (ed.). Perm: "Knizhnyy mir". (in Russian).

Surface water resources of the USSR. (1967). Basic hydrological characteristics. V. 11: Middle Urals and Cis-Urals. Issue 1. Ed. I.N. Moiseev. Leningrad: Gidrometeoizdat. (in Russian).

Tartakovsky, A.M. (ed.). (2012) *Atlas of the Perm Region*. Perm: Perm State National Research University. (in Russian).

Tolmachev, V.V. (1990). Classification of the consequences of damage to structures due to the impact of karst deformations. *Catastrophes and accidents in karst territories*, 9–10. (in Russian).

Yakovenko, P.I. (1969). *Sulfate karst of the Middle Urals and assessment of the stability of railway structures*: PhD thesis. Sverdlovsk. (in Russian).

Zaitsev, A.A., Pyankov, S.V., Bazhukova, N.V. (eds). (2020). *Geography Atlas of the Perm Region: study guide*. Perm: Perm State National Research University. (in Russian).

Zalkind, I.E., & Nechaev, Yu.A. (1959). *Limestone, dolomite and gypsum in the Perm region*. Perm: Perm publishing house. (in Russian).

3 Characteristics of Sulfate Karst Areas of the Perm Region

Photo by M. Maksimovich

The total area of regions where carbonate, sulfate, and salt types of karst are developed, according to the map compiled by K. A. Gorbunova, is 45.9 thousand km^2. The adjusted area according to our research is 47.5 thousand km^2. Thus, karst regions occupy almost a third of the Perm Region (160.6 thousand km^2).

In the Perm Region, karst is manifested in various geotectonic settings that determine geological conditions of its development: the composition of karst rocks, their relationship in the section, conditions of occurrence, degree of fracturing and complication by large ruptured disturbances, the presence of overburden, its composition. K. A. Gorbunova singled out groups of districts according to tectonic principle: karst regions of the eastern outskirts of the East European Platform and adjacent parts of the Cis-Ural trough, the Verkhnepechora depression, and Solikamsk depression of the Cis-Ural trough, Yuryuzan-Sylvin depression of the Cis-Ural trough, West Ural fold zone, and Central Ural uplift (Gorbunova et al. 1992).

Into the areas of sulfate karst, in this monograph, the authors include both territories where karsted are gypsum and anhydrite solely, as well as together with carbonate rocks and salts. According to the updated data, sulfate karst is distributed over an area of about 23 thousand km^2 or 15% Perm Region.

K. G. Butyrina, K. A. Gorbunova, G. A. Maksimovich, and others were involved in the zoning of sulfate karst in the Perm Region. In 1958, they identified the Polazninsko-Shalashninsky, Sylvinsko-Serginsky, and Kungursko-Irensky karst regions in the platform part of the Perm Cis-Urals, mainly in gypsum and anhydrite. These areas belong to the Ural karst province. Later, the zoning of the territory was revised (Gorbunova et al. 1992) and 16 karst regions were identified on the territory of the Perm Region.

According to the classification of Gorbunova et al. (1992), ten regions were identified where karst is associated with sulfate rocks. The authors of the monograph considered it possible to single out another—Lysvensky—area of local

N. Maksimovich et al., *Sulfate Karst of Perm Region (Russia)*, Cave and Karst Systems of the World,
https://doi.org/10.1007/978-3-031-18972-2_3

Table 3.1 Characteristics of sulfate karst regions of the Perm Region (Gorbunova et al. 1992 with additions by the authors)

Karst regions	Lithological type of karst	Area of the region, km^2	Karst type according to bedding conditions
1. Karst regions of the eastern margin of the East European platform and adjacent parts of the Cis-Ural trough			
Ksenofontovsky (Ks)	Sulfate, carbonate	1446	Bare, subeluvial, suballuvial, subfluvio-glacial, in some places—covered
Polaznensky (Pl)	Sulfate, carbonate-sulfate	3102	Covered, subeluvial, choked
Nizhnesylvinsky (Ns)	Sulfate, carbonate-sulfate	2191	Exposed, choked, subeluvial, suballuvial, covered
Irensky (Ir)	Sulfate, carbonate-sulfate	3815	Exposed, choked, subeluvial, suballuvial, covered
Kishertsky (Ksh)	Sulfate, carbonate-sulfate	334	Covered, suballuvial
2. Karst regions of the Upper Pechora depression and the Solikamsk depression of the Cis-Ural trough			
Visherksky (Vsh)	Sulfate, salt	3559	Subeluvial, suballuvial, subfluvioglacial, covered, in some places—exposed
Solikamsky (Sl)	Salt, sulfate-salt	7296	Covered, recovered
3. Karst areas of the Yuryuzan-Sylvinskaya depression of the Cis-Ural trough			
Osintsevsky (Os)	Sulfate, salt	785	Covered
Achitsky (Ach)	Sulfate	96	Covered
Kordonsky (Kr)	Sulfate	172	Covered
Lysvensky (Ls)	Sulfate	505	Covered, choked, in some places—exposed
Total area of karst regions		23,301	

distribution of sulfate karst, uniting the territories previously described by K. A. Gorbunova as areas of karst occurrences (Table 3.1; Fig. 3.1):

- Ksenofontovsky region of predominantly sulfate and carbonate karst;
- Visherksky region of salt and sulfate karst;
- Solikamsky region of predominantly salt and sulfate karst;
- Polaznensky region of predominantly sulfate and carbonate-sulfate karst;
- Nizhnesylvinsky region of sulfate and carbonate-sulfate karst;
- Kishertsky region of predominantly sulfate and carbonate-sulfate karst;
- Osintsevsky area of covered salt and sulfate karst;
- Kordonsky region of predominantly covered sulfate karst;
- Achitsky region of sulfate karst and klastokarst;
- Irensky region of predominantly sulfate and carbonate-sulfate karst;
- Lysvensky region of local distribution of sulfate karst.

With a detailed description of the areas which follows, a characteristic is given of the geological and hydrogeological conditions for the development of karst, karst phenomena associated with sulfate karst in specially protected natural areas, as well as areas that are proposed to be included in their composition. The influence of karst on economic activity and the ecological situation is described.

3.1 Ksenofontovsky Region of Predominantly Sulfate and Carbonate Karst

The region is located in the northern part of the Perm Region and occupies the basin of the middle and upper reaches of the Pilva River, a left tributary of the Kama River, between the Vizhaikha and the Bubyl Rivers, and further to the east —a site adjacent to the Kolva River, and covering an area of 1446 km^2 (Fig. 3.2). In the administrative aspect, it is located in the Cherdyn district, confined to the Ksenofontovo-Kolvinsky shaft (Gorbunova et al. 1992).

Geological conditions for the development of karst. The Kolvinskaya saddle is located in the northern part of the Cis-Ural trough. In the sublatitudinal direction, it joins the Timan Ridge with the Polyudovsko-Kolchimsky anticlinorium of the West Ural outer folding zone, and, in the meridional direction, it separates the Upper Pechersk and Solikamsk depressions of the Ural depression. The saddle is

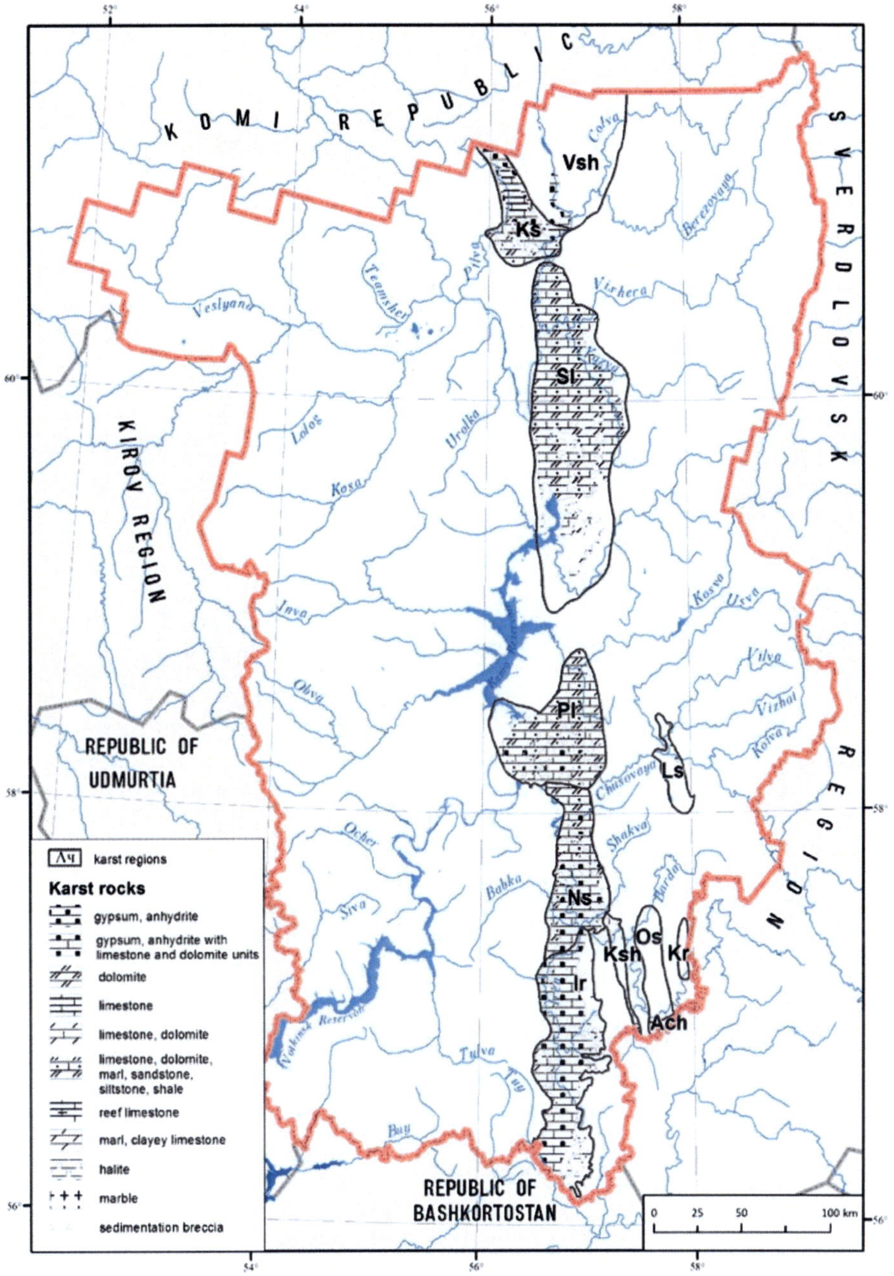

Fig. 3.1 Map of sulfate karst of the Perm Region (Gorbunova et al. 1992; with additions by the authors). *Note* Karst areas: Ks—Ksenofontovsky, Vsh—Visherksky, Sl—Solikamsky, Pl—Polaznensky, Ls—Lysvensky, Ns—Nizhnesylvinsky, Ksh—Kishertsky, Os—Osintsevsky, Ach—Achitsky, Kr—Kordonsky, Ir—Irensky

distinguished by an increased level of occurrence of deposits of the Lower Permian structural stage in comparison with their position in the boundary depressions.

In the vicinity of the Nyrob village, deposits of the Leksky and Koshelevsky formations of the Kungurian stage emerge, represented by sandstones, conglomerates, siltstones, and limestones with gypsum interlayers. Salt rocks of the Kama formation of the Ufimian stage come to the surface over a significant part of the territory. They are represented by platy limestones, dolomites, siltstones, sandstones, marls, and gypsum. The thickness of sulfate layers ranges from 20 to 180 m and carbonate layers from 2 to 10 m.

Hydrogeological conditions. The territory of the karst region belongs to the Ksenofontovo-Kolvinskaya hydrogeological region. The development of karst is mainly associated with the Kungurian carbonate-sulfate aquifer of the Irenian horizon.

Carbonate-sulfate rocks are water-bearing only in the upper part of the section in the places where they emerge to the surface and partially under conditions of occurrence

Fig. 3.2 Ksenofontovsky karst region

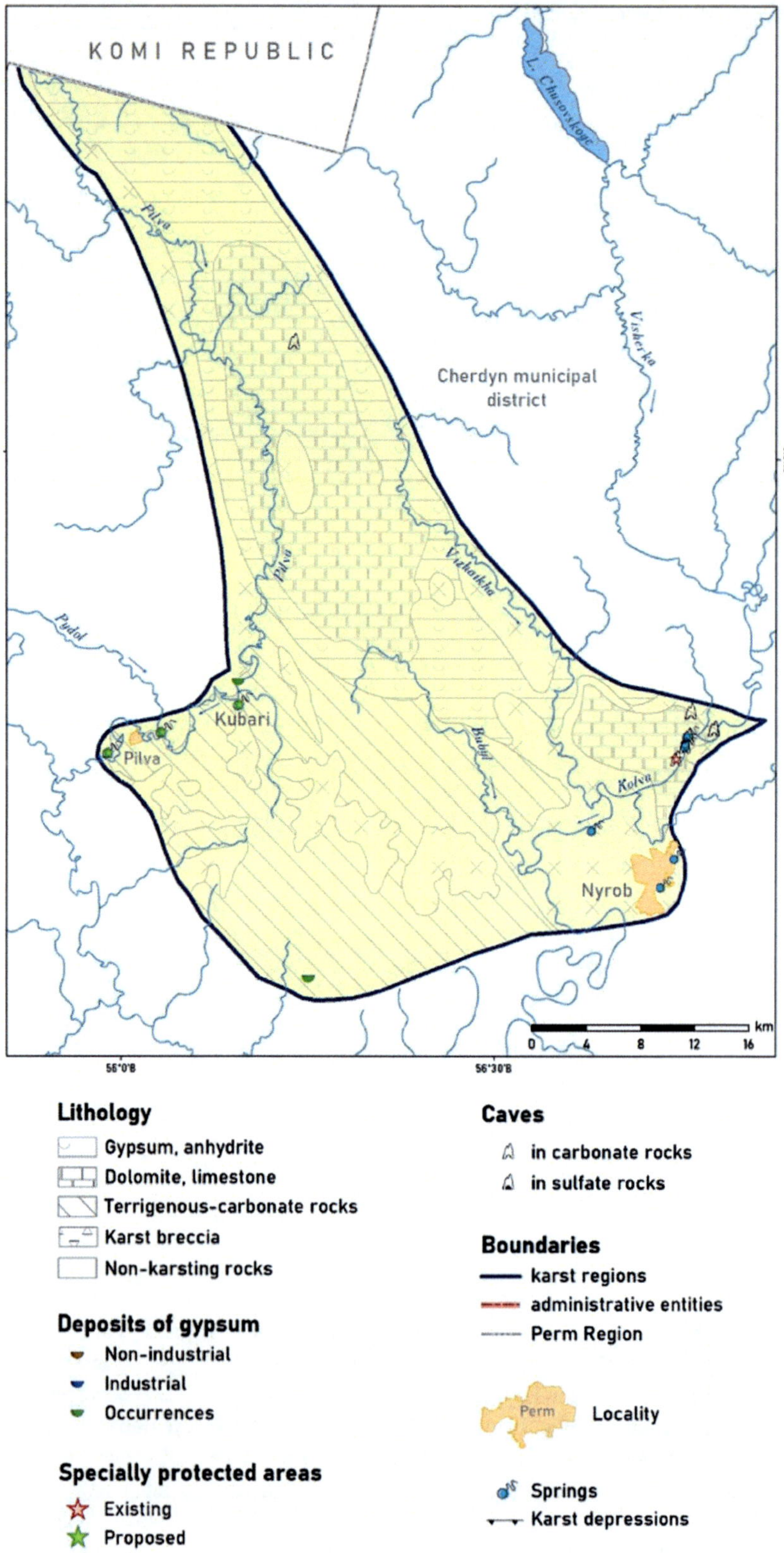

under thin Solikamsky rocks. With depth, fracturing decreases or becomes closed, and sulfates acquire the role of regional aquiclude. Along with heavily karst watered, monolithic impermeable rock blocks are also encountered. Groundwater flows through a network of disconnected karst cavities, channels, and cracks.

The main waterway of the region is the Kolva River, to the valley of which numerous springs are confined. Here, ground and fissure-vein gravity and pressure waters are widespread. The level of karst waters decreases in the direction of the rivers draining them: Pilva, Bubyl, Vizhaikha, etc. (Gorbunova et al. 1992).

The discharge of groundwater in the form of groups of sources in river valleys is confined to the areas where they intersect positive tectonic structures.

Groundwater discharging in the form of sources in the valley of the Kolva River belongs to various aquifers, formations, complexes, and zones. Some of them are located at the contact of aquiferous Solikamsky bedrock and locally aquiferous alluvial Quaternary deposits on the western boundary of the erosional-structural depression (springs Pokchinsky, Vilgortsky, and Paraskeva Pyatnitsa). The aquifers of this zone are limestones, marls, sandstones, and siltstones (Ikonnikov et al. 1965). Waters are fresh hydrocarbonate-calcium-sulfate with a mineralization of 240–430 mg/dm^3.

In the northeastern part of the described region (the Svyato-Nikolsky and Vilipsky springs), two zones of fresh water are distinguished: the southern zone of the Assel-Artinskian terrigenous complex and the northern weakly water-bearing suprazone of the Riphean-Lower Devonian terrigenous and metamorphic quartzite rocks. The waters of the Assel-Artinskian terrigenous complex, discharging in the Svyato-Nikolsky spring, belong to the hydrocarbonate-calcium-sulfate type with a mineralization of up to 360 mg/dm^3. The aquifers are interlayers and lenses of conglomerates, sandstones, less often siltstones, marls, and limestones. Also, this zone is associated with outcrops of hydrocarbonate-calcium-sulfate underground waters near the Divya Cave with a mineralization of up to 300 mg/dm^3. The waters of the weakly water-bearing suprazone of the Riphean-Lower Devonian terrigenous and metamorphic quartzite rocks are discharged in the Vilip spring and are characterized by hydrocarbonate-calcium-magnesium facies and mineralization up to 300 mg/dm^3.

Predominantly underground waters of the region belong to the same hydrocarbonate-calcium-sulfate hydrochemical facies, except for the Vilipsky spring, which has a hydrocarbonate-calcium-magnesium composition. The facies composition of the waters of the springs of Pokchinsky and Paraskeva Pyatnitsa changes in the summer-autumn period from hydrocarbonate-calcium-sulfate to hydrocarbonate-calcium-magnesium and back in the winter-spring period.

Below the main drains and under conditions of occurrence of the Kungurian carbonate-sulfate complex under younger sediments, the intensity of karst decreases, which is associated with a decrease in the velocity of groundwater movement and saturation with sulfates (Gorbunova et al. 1992).

On the territory of the Pre-Timan trough, outlets of hydrogen sulfide waters of sulfate–calcium composition are known in the valleys of the Pydol and Pilva Rivers. They are confined mainly to the deposits of the Philippovsky horizon. Hydrogen sulfide waters of sodium chloride composition were exposed in these sediments (Fig. 3.3).

The Pydolsky spring is located 8 km northwest of the Kubari village, on the left bank of the Pydol River, right tributary of the Pilva River, which flows into the Kama River. Here, 5 km from the mouth of the Pydol River in the swampy lowland of the river floodplain, in a dense, impassable forest 50 m from the river bank, powerful hydrogen sulfide springs emerge, giving rise to the Usolka River. At the mouth of the spring, there is a wooden pipe 28 cm in diameter. At 10 m from the pipe, there is a deep ponor resembling a karst well, clearly visible to a depth of 6–7 m, along which a powerful stream of hydrogen sulfide water also rises.

The spring water is clear, with a bluish tint in ponor. Along the entire channel at the bottom of the Usolka River on pebbles and grass, there are numerous white deposits of sulfur (barenjima). The spring water belongs to the group of low-mineralized waters of sulfate–calcium-hydrocarbonate hydrochemical facies. The content of hydrogen sulfide is more than 100 mg/dm^3.

The Uzhginsky spring (Lake Koch) has a horseshoe shape, 250 m long and 35–40 m wide. The lake bottom is viscous, covered with thin brownish silty mud, which

Fig. 3.3 Valley of the Pilva River near the western border of the Ksenofontovsky karst region (*photo by* N. Maksimovich)

contains 36–42 mg/dm^3 of hydrogen sulfide and 690–725.8 mg/dm^3 carbonate carbon dioxide. The content of inorganic carbon dioxide reaches 4752 mg/dm^3, which indicates the active decomposition of organic substances. In the water extract, ions of bicarbonates, calcium, and sodium sulfates predominate, and the pomace belongs to the sulfate-hydrocarbonate-chloride-calcium hydrochemical facies and to the low-mineralized group.

The water in the lake is clear, bluish-green in color, fresh in taste with a characteristic harsh aftertaste and a faint smell of hydrogen sulfide. According to the chemical composition, the water of Lake Koch belongs to the group of low-mineralized waters of the sulfate–calcium hydrochemical facies. Spectral analysis in the dry residue established the presence of strontium (4.38 mg/dm^3), boron (0.09 mg/dm^3 HBO_2), and other trace elements. The river flowing out of Lake Koch is about 200 m long and flows into the Pilva River. The chemical composition of the water almost does not differ from Lake Koch.

Pilva spring. During the drilling of structural prospecting wells in the Pilvenskaya square (the Kubari village neighborhood), pressure hydrogen sulfide waters of sodium chloride composition were discovered among the Artinskian and Philippovsky deposits. The chemical composition of waters has been studied very poorly. The selected water samples on the self-flow are represented by a mixture of overlying fresh aquifers. The most saline water was obtained from a self-flowing well from a depth of 392 m from the Philippov dolomites of the Kungurian stage. The water had a bitter-salty taste with a pungent smell of hydrogen sulfide. In terms of chemical composition, it belongs to the group of moderately mineralized waters of the sodium chloride hydrochemical facies with increased contents of sulfate and calcium ions. The underground waters of the overlying Solikamsky deposits are represented by fissure-karst waters of sulfate–calcium composition.

Karst phenomena. Karst is developed mainly in carbonate-sulfate rocks of the Kungurian and Ufimian stages. In watersheds, sulfate deposits, exposed by erosion and located in the zone of active water exchange, are subject to intense karsting (Gorbunova et al. 1992). To date, there is no information about karst forms and caves in sulfate deposits due to the inaccessibility of the area.

Existing and proposed specially protected natural areas (SPNA)

Existing SPNAs. The Ksenofontovsky karst region is poorly studied in terms of karstology. To identify unique objects associated with the manifestation of sulfate karst, which can be attributed to SPNA, it is necessary to organize an expeditionary survey of the Ksenofontovo district. Currently, there is one SPNA of regional significance associated with the manifestation of karst.

Fig. 3.4 Diviy Kamen Spring (*photo by* A. Kazantseva)

The landscape natural Monument Diviy Kamen and the Cave has an area of 375 ha and is located on the right bank of the Kolva River, 10 km north of the Nyrob village, Cherdyn district. Divya is the longest cave in the Urals, its length is 10.1 km. Proposed for protection by N. P. Yastrebov in 1958. The entrance is located in a forest at an altitude of 120 m above the river's edge on a steep coastal slope called Diviy Kamen. At an altitude of 140 m, it turns into a watershed plateau.

The cave is developed by gently dipping to the north of the Lower Permian layered silicified limestones of the Sakmarian and Artinskian stages. In the lower part of the cave, light gray layered limestones with the inclusion of limonitized pyrite were found, passing higher into limestones with the fauna of foraminifera, mollusks, bryozoans, rugoses, and crinoids. On the territory of the landscape monument, unloading of hydrocarbonate-calcium-sulfate underground waters with a salinity of up to 300 mg/dm^3 takes place (Fig. 3.4).

Proposed SPNAs. Promising objects for the organization of SPNAs are the Pydolsk springs and Lake Koch, described above.

Economic activity and environmental pressure. The basis of the region's industry is made up of logging and timber processing industries. The technogenic load on karst rocks is insignificant.

3.2 Visherksky Region of Salt and Sulfate Karst

The region has an area of 3559 km^2, is located within the Cherdyn district of the Perm Region, and occupies the southern part of the Verkhnepechora depression in the basins of the lower reaches of the Kolva River and its tributary—the Visherka River (Fig. 3.5).

Fig. 3.5 Visherksky karst region

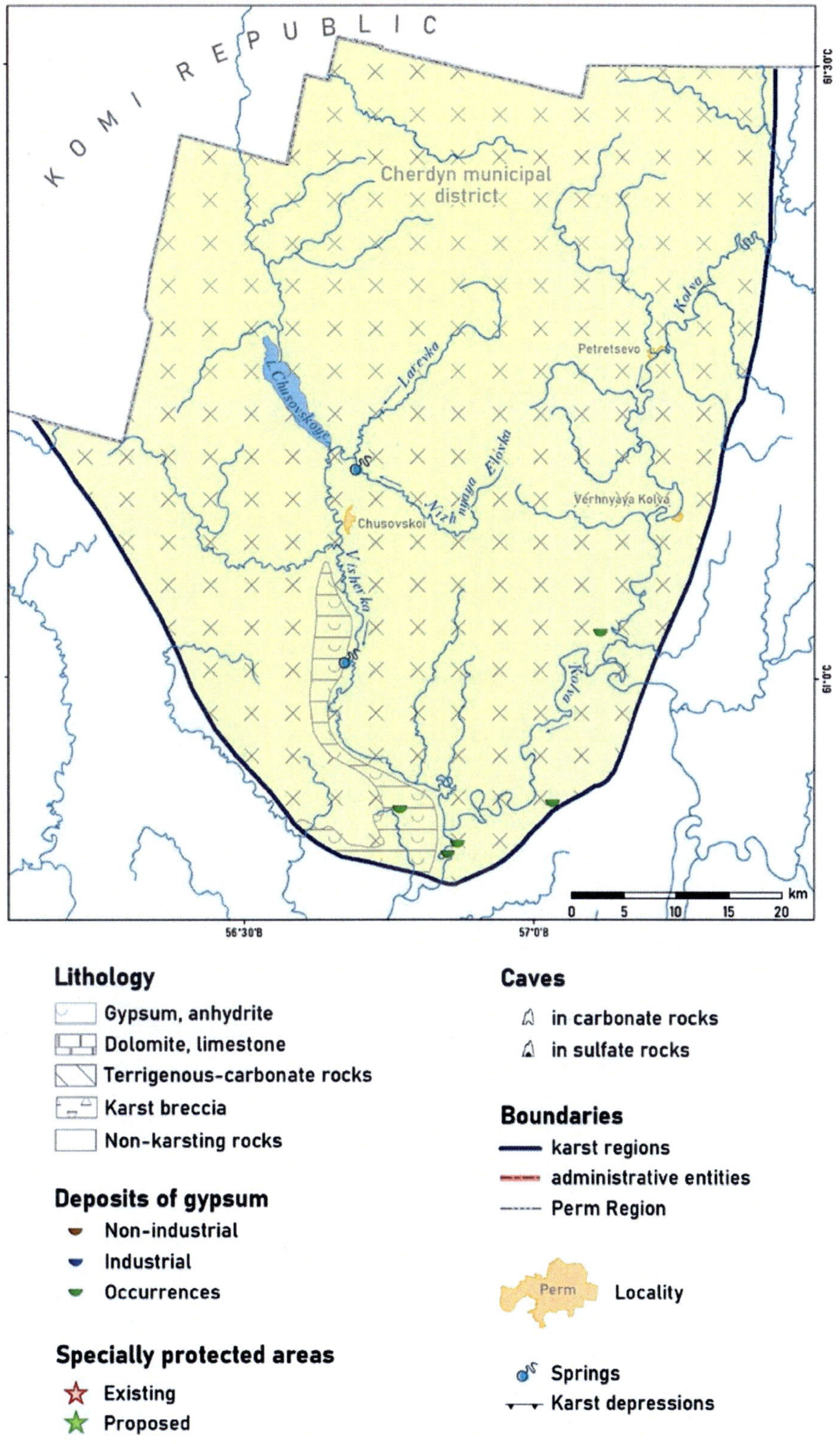

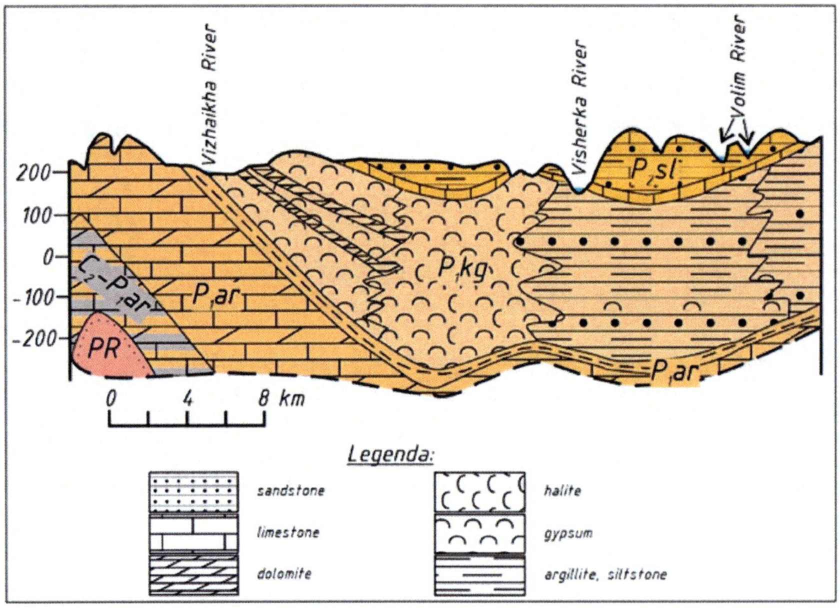

Fig. 3.6 Geological conditions for the development of karst in the Visherksky karst region (Gorbunova et al. 1992)

Geological conditions for the development of karst. Tectonically, the region is located within the Verkhnepechora depression. The eastern border of the region is complicated by the main Patrokovsky and Kikusky thrusts. In terms of geomorphological zoning, the territory belongs to the denudation region of the Russian Plain.

Lower Permian deposits are subdivided into two parts: lower—carbonate and upper terrigenous (Fig. 3.6). The boundary between them runs inside the Artinskian stage and is well recognized in aerial photographs. The Lower Permian carbonate stratum includes reef and bedded facies (Snitko et al. 2015).

Terrigenous sediments of the Artinskian stage in the west begin with a stratum of clay shales and marls up to 40–50 m thick. Above, there is a sandstone-shale stratum, the thickness of which reaches 100 m in places, and, in some places, it almost completely disappears. Sporadically appearing conglomerate interlayers are confined to the very top of this formation. Even higher, there is a new sandy-marl formation, represented by a rather thick stratum of greenish-gray marly clays and marls with subordinate interlayers of fine-grained sandstones and conglomerates. In a number of sections, large lenticular stock-like gypsum deposits are confined to the base of this formation (the right bank of the Valyasherka River, the White Mkhi tract, the vicinity of the Kikus village, the vicinity of the Raskat village, etc.) (Fig. 3.7). Obviously, in addition to gypsum, small deposits of rock salt also belong to the bottom of this stratum, with which salt springs on the Usolka River, right tributary of the Visherka River, are connected (Kalmykova and Dutkevich 1934).

Fig. 3.7 Outcrop with gypsum development near the Bolshoi Kikus village (*photo by* O. Kadebskaya)

The most ancient terrigenous deposits of the Artinskian stage come to the surface at the border with the folded zone of the Urals. In the submerged axial part of the depression, there is a thick (more than 350 m) stratum of gypsum, anhydrite, salt interbedded with dolomites, and gypsum terrigenous rocks of the Kungurian stage. Gypsum and anhydrite come to the surface in the valleys of the Kolva and Visherka Rivers. The overlying terrigenous stratum of the Solikamsky horizon of the Ufimian stage has different thicknesses depending on the relief. It is distributed in the northern and northwestern parts of the region. The cover sediments belong to the eluvial-deluvial type in the watersheds, and the alluvial type in the river valleys (Gorbunova et al. 1992).

Hydrogeological conditions. The Visherksky karst region belongs to the Pechora-Pre-Ural basin of stratal and block-stratal pressure waters of the Verkhnepechora hydrogeological region. Ground waters of alluvial and fluvioglacial sediments, Solikamsky and Kungurian terrigenous aquifers and fissure-karst waters of halogen rocks of the Kungurian stage are widespread here. Salinity and gypsum content of the Kungurian and, to a lesser extent, Solikamsky deposits determine the shallow bedding of saline waters. The valleys of the Visherka (from the Larevka River to the Fadino tract) and the Kolva Rivers (from the Petretsovo to the Kikus villages) are discharge zones of salt and brackish waters of chloride and chloride-sulfate composition (Gorbunova et al. 1992; Snitko et al. 2015).

The aquiferous Kungurian sulfate-terrigenous complex is represented by terrigenous deposits of the Kungurian stage. Sandstones, siltstones, and conglomerates are aquifers; confining stratum—non-fractured differences of these rocks, clay, gypsum, and anhydrite. The filtration properties of the complex depend on the lithology of the water-bearing rocks and the degree of their fracturing and are 0.001–14.7 m per day. In the upper part of the section, fractured groundwater is developed, and the depth of the level varies from 12 to 52 m. Above the erosional incision, hydrocarbonate and sulfate-hydrocarbonate waters with a salinity of 0.1–0.5 g/dm^3 prevail. Sulfate ions predominate in the areas of gypsum and anhydrite leaching. Mineralization of these waters is 0.3–3.0 g/dm^3. The main source of groundwater supply is precipitation. The natural protection of groundwater from pollution is characterized as low and medium (Snitko et al. 2015).

Karst phenomena. Karst belongs to the sulfate type in gypsum, anhydrite, gypsum sandstones, and salt type in the salt-bearing stratum of the Kungurian stage. Subeluvial, suballuvial, subfluvioglacial, and covered types of karst prevail, in some places bare karst occurs. Sulfate karst is intensely manifested in the form of fields of large craters in areas where sandy-argillaceous deposits, including gypsum, are developed.

The valleys of the rivers flowing through the lowland are swampy. The river flow is calm, which is typical for the areas of salt karst.

Apparently, the close location of the groundwater recharge area—Timan—has a significant impact on the development of karst. From its territory, low-mineralized waters are supplied to the sediments of the Ufimian and Kungurian stages, which are capable of dissolving sulfate-halogen rocks in significant quantities. The presence of salts increases the solubility of sulfates tenfold. The dissolution of sulfate-halogen rocks is evidenced by the chloride-sodium-sulfate composition of the water of springs, lakes, rivers, and wells (Fig. 3.8). Mineralization of groundwater and some lakes ranges from 5 to 34 g/dm^3.

Lake Chusovskoye is located in a vast erosional-karst lowland about 40 km long and 2–7 km wide. According to I. N. Shestov and A. V. Shurubor, the lake was formed as a result of salt-gypsum karst, which is confirmed by the occurrence of salts, gypsum, and anhydrite in the zone of active water exchange (Gorbunova et al. 1992).

Surface karst forms are developed: craters, lakes, depressions, and ravines. Among the karst sinkholes, corrosion and corrosion-erosion types prevail. In the area of the Petretsovo village, as a result of leaching of gypsum and anhydrite and the subsequent collapse of the roof of the

a

b

Fig. 3.8 Karst springs (*photo by* O. Kadebskaya): **a**—near the Maly Kikus village; **b**—near the Bolshoi Kikus village overlying terrigenous rocks, dips up to 300 m in diameter were formed (Snitko et al. 2015).

Proposed specially protected natural areas. There are no existing SPNAs on the territory of this region. In our opinion, it is necessary to highlight the manifestations of highly mineralized waters and mud—Larevsky springs and lakes. They are located in the lower reaches of the Nizhnyaya Elovka River, on the left bank, 800 m from the mouth or 2.5 km southeast of the Larevka village. Salt springs are common in the area called by the local population "Maryeva poyma". Here, on an area of about 8 ha, swampy solonetzes are developed, represented by black viscous mud with salt efflorescence on the surface and with a pungent smell of hydrogen sulfide. The surface of the salt licks is covered with a mud-brown crust with drying cracks. The outlet of the springs is represented by small griffins. The water of the springs is cloudy with a faint smell of hydrogen sulfide and a bitter-salty taste. In terms of chemical composition, it belongs to the group of highly mineralized waters of the sodium chloride hydrochemical facies.

To the northwest of the Larevskiy spring, there is a group of salt lakes, which are fed by the discharge of fissure-karst waters of gypsum-salt karst and atmospheric precipitation (Geological monuments … 2009).

Economic activity and environmental pressure. The basis of the region's industry is represented by logging and timber processing industries (Fig. 3.9). It is possible that logging, which is accompanied by a change in surface runoff, infiltration, etc., affects the activity of karst processes.

On March 23, 1971, in the area of Lake Chusovskoye, as part of a project to transfer northern rivers in order to create an artificial canal, three underground nuclear explosions with a capacity of 45 kilotons at a depth of 127 m were carried out. After the explosion, a gas-dust cloud with a height of 1800 m was created. A lake 750 m long, 350 m wide, and 12–14 m deep was formed at the site of the explosion. The lake has no tributaries and is not connected to the river network. Unfortunately, the effect of explosions on changes in hydrogeological conditions and the activity of the karst process has not been studied.

Fig. 3.9 Burning peat on the banks of the Kolva River at the confluence of the Visherka River (*photo by* O. Kadebskaya)

3.3 Solikamsky Region of Predominantly Salt and Sulfate Karst

The area of the Solikamsky karst region is 7296 km^2. Its eastern border in the north coincides with the border of the folded Urals, to the south it can be traced along the right bank of the Glukhaya Vilva River and the valley of the Usolka River (Gorbunova et al. 1992) (Fig. 3.10). The area extends from the lower reaches of the Nizva River to the south to the interfluve of the Yayva and the Kosva Rivers. From the west, it is bounded by the meridional sections of the Kolva and the Vishera Rivers, and to the south by the valley of the Kama reservoir. Administratively, it is confined to the territory of the Alexandrovsk, Berezniki, Cherdyn, Dobryanka, Krasnovishersk, and Solikamsk districts of the Perm Region.

Geomorphologically, the region is a vast denudation plain formed by numerous vaulted structures and depressions separating them. The denudation plain is dissected by a rather dense network of small river valleys, gullies, and ravines. The rivers of the region belong mainly to the Kama River basin.

Geological conditions for the development of karst. The region is confined to the Solikamsk depression, complicated by the Bereznikovsky and Unvinsky ledges (Fig. 3.11).

The Verkhnekamskoye deposit of potassium-magnesium salts (VDPMS) is associated with the salt rocks of the depression. The lens-shaped salt stratum is a multi-layer deposit of potassium-magnesium and sodium salts, where the underlying rock salt, sylvinite and carnallite zones, and overlying rock salt replace each other from the bottom up. Salt deposits belong to the Irenian horizon of the Kungurian stage. Higher in the section, with angular unconformity, there are deposits of the Solikamsky (salt-marl [SMS] and terrigenous-carbonate stratum [TCS]) and Sheshminsky (variegated stratum [VGS]) horizons of the Ufimian stage. The structure of the field is disturbed by sublatitudinal faults, bordering from the south the Durinskaya and Borovitskaya depressions, which cut into a salt deposit (Tchaikovsky 2008).

The underlying rock salt is divided into three packs. The lower layer is composed of alternating layers of rock salt and terrigenous-chemogenic rocks (argillaceous-(marl)-anhydrite-dolomite, argillaceous-(marl)-gypsum-anhydrite, and dolomite-anhydrite varieties, as well as siltstones and sandstones). The middle layer is represented by a thick homogeneous rock salt with interlayers of clay-anhydrite

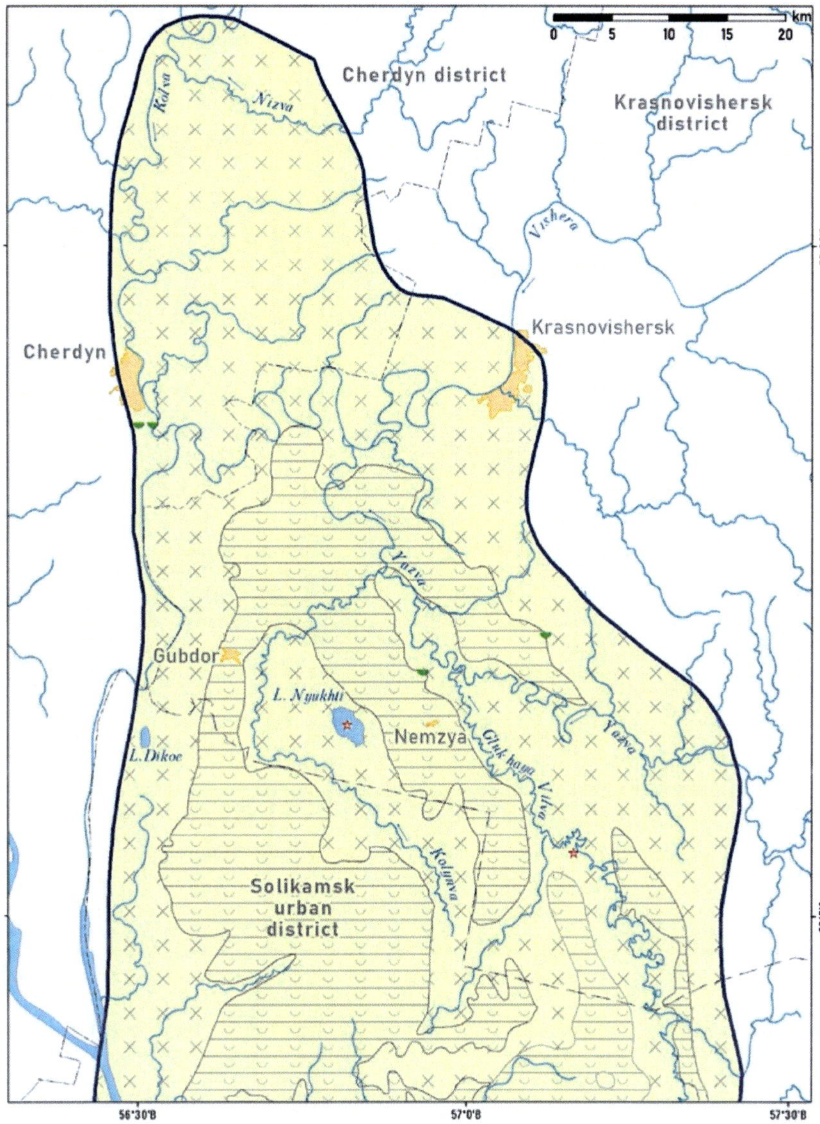

Fig. 3.10 Solikamsky karst region

material from 0.05 to 0.5 m thick. The upper layer is represented by rock salt (Kudryashov 2013).

Covering rock salt (CRS) is widespread in the dominant area of the VDPMS, but is absent on the arches of some uplifts.

Rock salt is gray and light gray, in places light yellow and pale pink; in the lower part, disseminated carnallite and sylvite are noted. The rock has a pronounced layered texture, due to the alternation of interlayers of various structural varieties of halite and clay-anhydrite material.

The Solikamsky horizon is identified in the volume of the Solikamsky formation, which is divided into two subformations.

The Lower Solikamsky subformation, crowning the section of the halogen formation, is represented by the SMS. The stratum is composed of marls, clays, rock salt, and, to a lesser extent, limestones, with interlayers of gypsum, clay-gypsum-carbonate layers with anhydrite nodules. Low permeability layers: clays and gypsum, determine the overall low permeability of the stratum in the vertical direction (Potapov 2016). The average composition of the stratum section: marls and clays—65%, rock salt—30%, gypsum and anhydrite—5% (Kudryashov 2013) (Fig. 3.12). The clayey-carbonate rocks of the SMS often contain pyrite, coalified plant remains, and the fauna of pelecypods. The section of the Solikamsky suite of the supra-salt stratum in the Polovodovsky area (along the borehole 704/1) and the forms of gypsum are shown in Fig. 3.13.

Similar formations were recorded on one of the sides of the Durinsky trough. Some researchers associate their

Fig. 3.10 (continued)

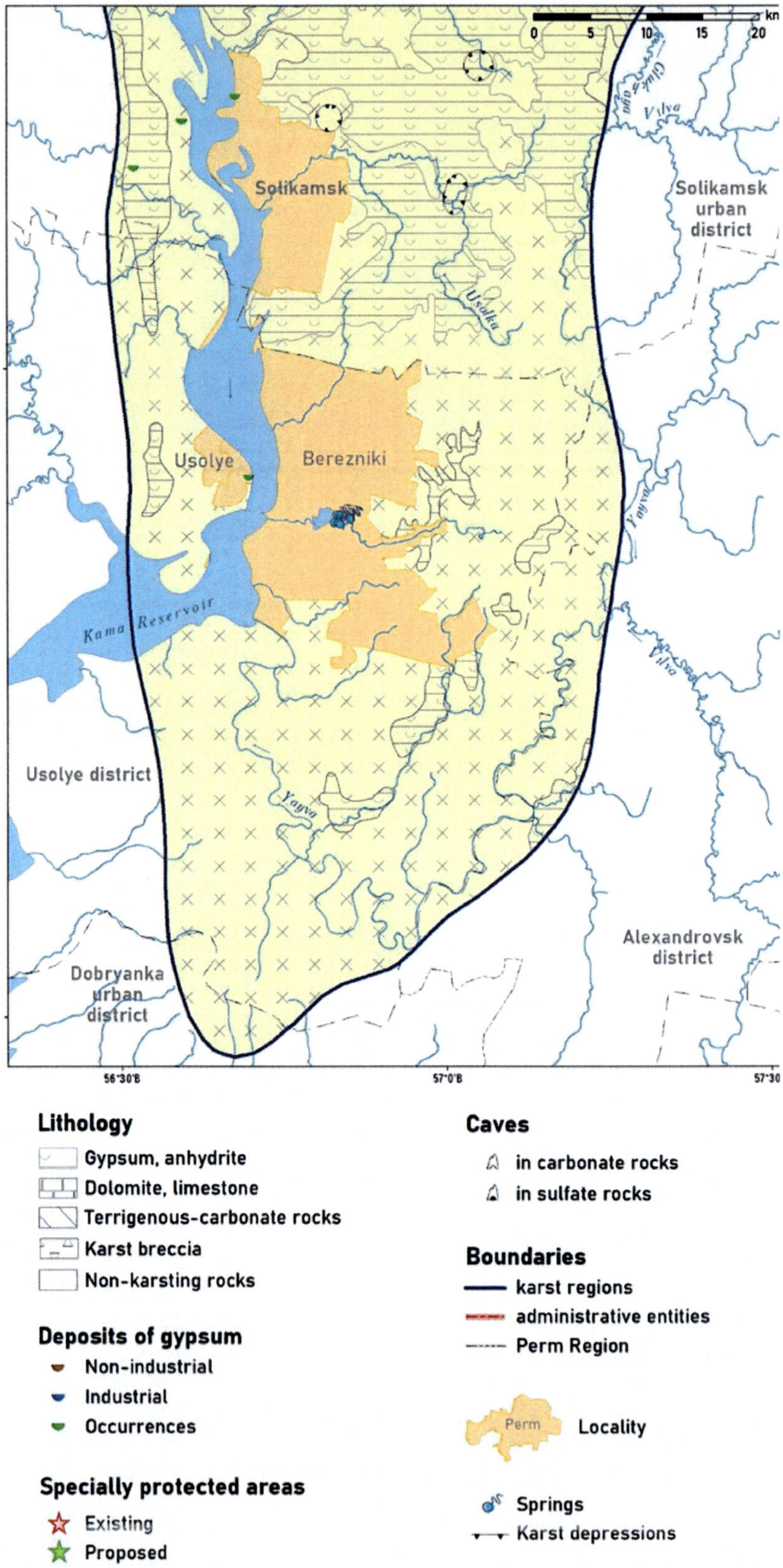

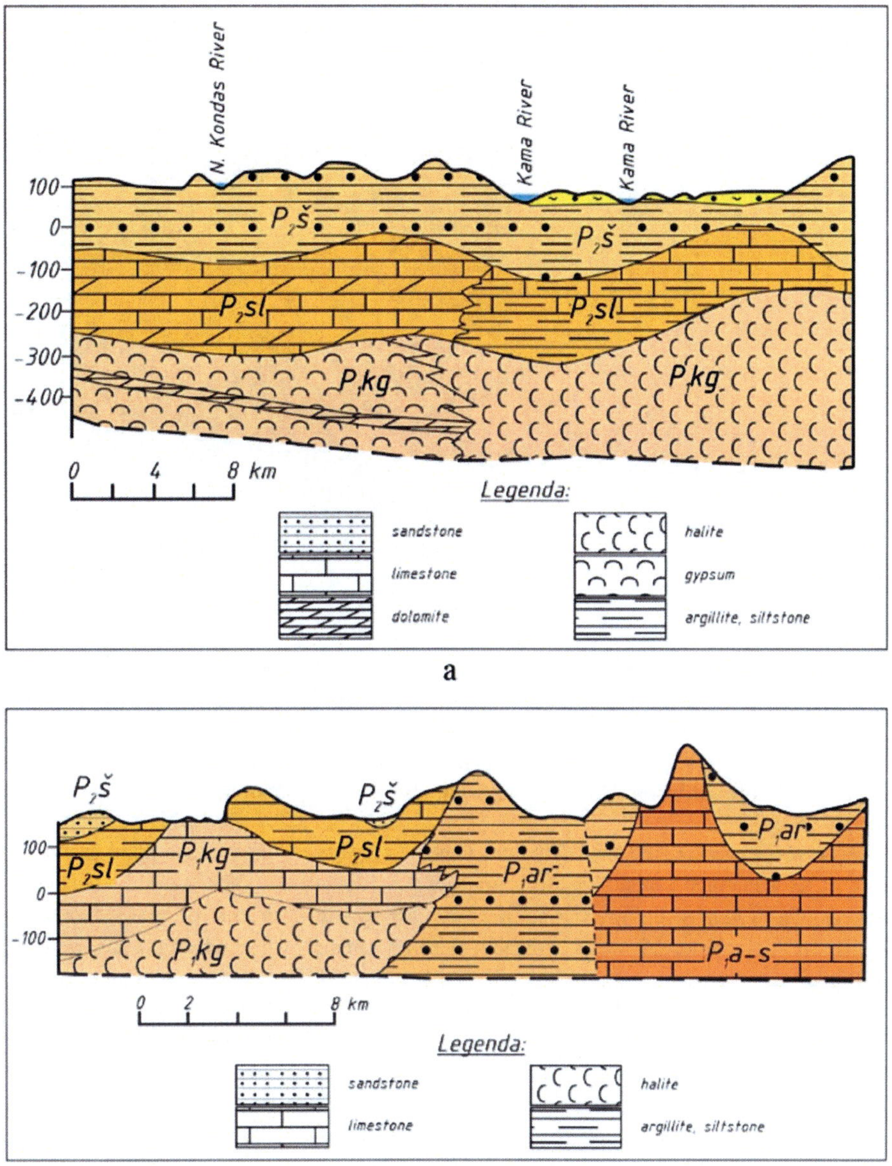

b

Fig. 3.11 Geological conditions of karst development in Solikamsky region: **a**—western part; **b**—eastern part (Gorbunova et al. 1992)

formation with karst provoked by near-fault fracturing, but as a result of examining the core of one of the boreholes, it turned out that the main role belongs not to karst processes, but to the tectonic factor (Kalinina and Tchaikovsky 2014).

Gypsum-clay hats are examples of the lithological transformation of a karst substrate—a hypergene replacement of its karst part by weakly or insoluble formations, which occurs as a result of prolonged karstification. The degree of material transformation can reach 10–30% or more (Andrejczuk 2007).

The Upper Solikamsky subformation is represented by the TCS. The sequence contains limestones, dolomites, thin-layered marls, mudstones, siltstones, fine-grained sandstones, and limestones. All TCS rocks are fractured to one degree or another. The thickness of the stratum is 80 m on average.

Above, the Sheshminsky formation occurs, represented by the VGS. The sequence is composed of brown, greenish-gray, and gray sandstones and siltstones, sometimes with thin layers of marls and limestones. The average thickness of the sediments is 25 m. There is no VGS in some areas. Plastering is observed in the form of lenticular interlayers, consonants, and intersecting gypsum veins. Within the deposit, the thickness of the Sheshminsky horizon varies from 0 to 675 m. The maximum thickness is confined to the Durinskaya area of the deposit (Kudryashov 2013).

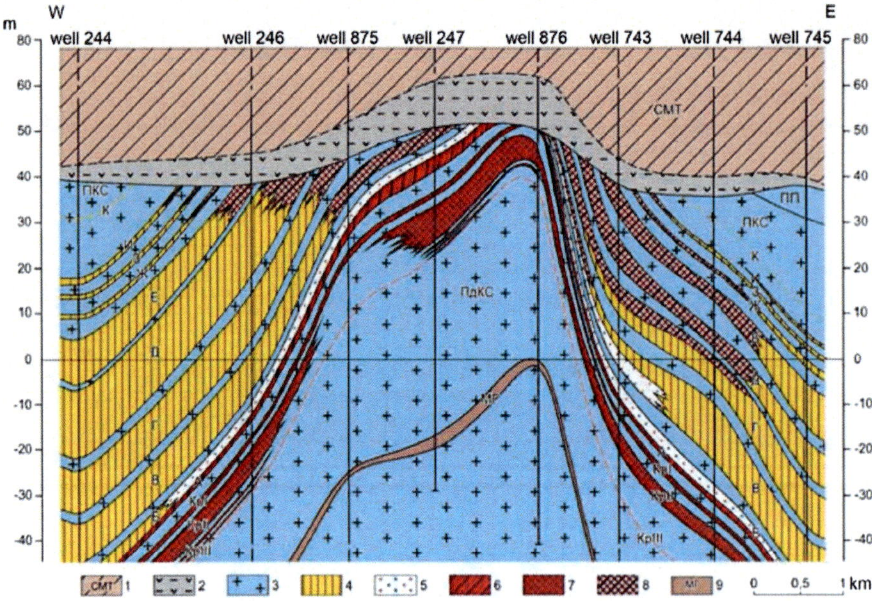

Fig. 3.12 Geological section of the upper part of the salt stratum on the Kharyushinsky uplift (Kudryashov 2013). 1—rocks of the salt-marl stratum; 2—rocks of a plaster-clay hat; 3—rock salt; 4—carnallite rock; 5—variegated sylvinite; 6—banded sylvinite; 7—red sylvinite; 8—hypergenic sylvinite; 9—clay deposits

Deposits of the Quaternary system are represented by clays, loams, sandy loams, sands, pebbles, as well as peat-swamp deposits. By their genesis, the deposits are classified as eluvial-diluvial, eolian, fluvioglacial, lacustrine-boggy, alluvial, and technogenic (Gorbunova et al. 1992).

Hydrogeological conditions. The salt deposit plays the role of an aquiclude separating the supra-salt and sub-salt waters. The waters of the pre-salt complex are highly mineralized brines. Supra-salt waters are both fresh and mineralized. The supra-salt stratum contains several aquifers. The Quaternary aquifer is confined to alluvial, lacustrine, fluvioglacial, and eluvial-diluvial sediments. These are non-pressurized groundwaters (Gorbunova et al. 1992).

The main reserves of fresh groundwater in the region are concentrated in the deposits of the TCS. The water-bearing rocks are limestones, marls, and sandstones. Clays and argillaceous siltstones are used as seals. The TCS aquifer in the Solikamsky karst region has a thickness of up to 100–135 m. The complex distinguishes between two types of groundwater: fractured groundwater, usually unconfined, which move from watersheds to drainage valleys and are located above the local drainage base; fractured-stratal, the movement of which from the recharge area is subordinated to the general direction of the layers falling to the west to the axis of the Kama trough.

Fractured reservoir waters of the region are pressure-bearing, as evidenced by the presence of ascending springs, self-flowing from boreholes. Under natural conditions, the underground waters of the TCS have a hydrocarbonate-calcium composition and a total mineralization of 0.2–0.5 g per dm^3. The flow rates of the springs reach 20 L per second.

The SMS aquifer (the Upper and Lower Solikamsky aquifers) is located in fissured marls, where clay interlayers are aquicludes. The main basis for the drainage of the region is the Kama River valley, into which the discharge of the pressurized underground waters of the horizon is carried out.

The natural chemical composition and salinity of the waters of the SMS aquifer are very diverse. Fresh underground waters are widespread in places where SMS deposits emerge on the surface. The sediments of the Lower Solikamsky horizon of the SMT, overlain by the Upper Solikamsky rocks, are located below the drainage zone and contain brackish groundwater. Leaching of the gypsum clays present in the SMS section results in the presence of brackish sulfate–calcium waters. In areas where gypsum clays are shallow, sulfate waters appear at shallow depths.

The aquifer of the middle-lower layers of the SMS lies at a depth of 200–240 m, circulates near the salt table, and has a mineralization of 120–300 g/dm^3.

The reservoir properties of the horizons of the supra-salt aquifer are mainly due to fracturing and cavernousness formed in the course of rock hypergenesis. Salt and gypsum mirrors are the main surfaces reflecting the depth of multi-temporal hypergene transformations of rocks and, as a consequence, their hydrogeological stratification. The salt surface does not represent a stratigraphic surface but serves as the boundary between water-resistant saline and aquiferous supra-salt rocks. The gypsum mirror, located in most of the territory in the lower part of the TCS, stands out along the first interlayers and inclusions of gypsum from the

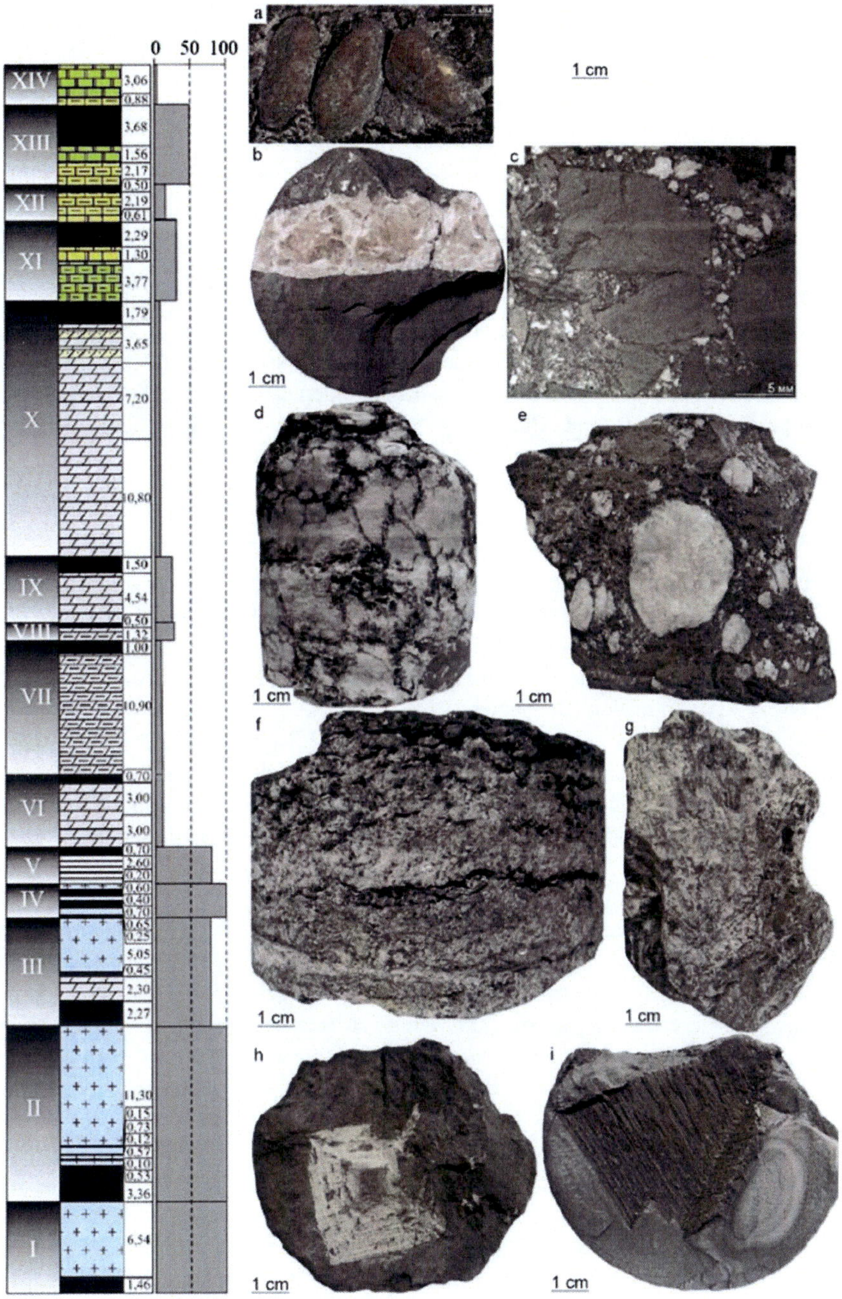

Fig. 3.13 Lithological section of the Solikamsky formation of the post-salt stratum of the VDPMS along the borehole 704/1 (Polovodovsky area) (Kalinina 2015). *Notes* to Fig. 3.13: **a**—bivalve shells; **b**–**f**—gypsum nodules (**b**—in cutting cracks; **c**—in the composition of cement breccia; **d**, **e**—interlayers; **f**—massive gypsum interlayer); **g**—an interlayer of gypsum with a "shadow" chevron structure; **h**, **i**—pseudomorphs along the "boats" of halite

surface and separates the zone of active water exchange with fresh and slightly saline waters from the zone of difficult water exchange with saline waters and brines up to 100–150 g/dm^3. Sulfate-bearing rocks in the upper layers are universally affected by sulfate leaching processes. Blocks of weakly fractured and non-cavernous gypsum-bearing rocks located below the gypsum mirror are relatively water-resistant. On the territory of the VDPMS, the surfaces of the salt and gypsum mirrors are separated by the depth of the supra-salt section by tens of meters (Potapov 2016).

Accumulations of highly mineralized brines are found not only at contact with salts but also somewhat higher (the first tens of meters), in beds and layers of karst gypsum of SMS. In these cases, brine-hosting rocks are usually represented by two or three (rarely more) thin, contiguous layers (Kudryashov 2013).

According to I. I. Tchaikovsky (2018), at the Yuzhno-Solikamsk area of the VDPMS in 2017, a leaching zone was opened, in which the salt layer was replaced by cavernous gypsum-syngenite-halite rock with unusual brines: super-hard (354.68–375, 12 g/dm^3), chloride potassium-sodium composition (*Na* 97.18–105.50; *K* 47.65–53.00; *Mg* 0.48–1.26; *Ca* 1.09–1.61; *Cl* 199; 51–211.12; SO_4 4.18–4.76; *Br* 0.21–0.34; HCO_3 0–0.12; CO_3 0–0.07 g/dm^3), pH 6.0–7, 1. The encountered brines are comparable to the waters in contact with rock salt (brine horizon); therefore, it can be assumed that they came from the surface of the salt strata and have paleokarst nature. Sulfate rocks were most likely the source of sulfate ions.

Karst phenomena. The Solikamsky karst region differs from the rest of the regions by the originality of the conditions of karst formation and types of karst. Salt, sulfate-salt karst is developed here. Karsting rocks occur underneath non-karsting bedrocks and Quaternary sediments. Karst activation and suffusion processes occur mainly as a result of human engineering and technical activities (Gorbunova et al. 1992).

Several types of karst are distinguished in the region: natural, which has arisen due to the presence of the basic conditions for the development of karst; technogenic, caused by human engineering and economic activities; mixed, developing under the influence of one or two artificial conditions of karst formation (Beltyukov 1978).

Modern and ancient salt and, to a lesser extent, sulfate karst appear on the surface during the formation of subsidence troughs, extended areas of valleys with a large thickness of alluvial, alluvial-lacustrine and, in some places, fluvioglacial deposits, and lacustrine basins (Gorbunova et al. 1992).

On the territory of the VDPMS, there are two zones of hypergene karst. Zones of local karst formation are located on the arches of positive structures; however, karst development is observed mainly on structures whose surfaces are associated with streams. Zones of continuous development of karst are located in an area with a general high occurrence of salt rocks, excluding areas where the salt surface decreases. There is a well-developed hydrographic network on the surface, the area is watered and swampy.

Hypergene karst is the reason for the appearance of areas of the first type of local variability of the mineral composition of the productive strata. Meteoric water, when it comes into contact with salt rocks, transforms them. Analysis of lithological maps in the places of occurrence of karst showed that this local variability is characterized by zoning: carnallite rocks → sylvinites → rock salt → rocks of gypsum-clay hats (Kudryashov 2013).

The most extensive karst depressions with a diameter of up to 10–15 km are located in the northern part of the Solikamsk depression near the settlements of Kuznetsovo, Gubdor, and Nemzya (Dorofeev and Lukin 1969).

The flat bottoms of the depressions are swampy. The Vishera, Kolynva, Glukhaya Vilva Rivers flow along them.

Fig. 3.14 Lake Nyukhti located at the bottom of a karst depression (*photo by* N. Maksimovich)

Among the bogs, there are numerous shallow lakes, the largest of which are Nyukhti (5.5 km^2) (Fig. 3.14) and Dikoe (1.5 km^2). The dividing elevations of the submeridional strike rise up to 50–70 m above the bottoms of the depressions.

The formation of negative landforms is mainly associated with the dissolution of rock and potassium-magnesium salts with a much lesser role of gypsum and anhydrite. Surface manifestations of pure sulfate karst are rarely found in the region. In the valley of the Yazva River, deposits of the Koshelevsky suite come to the surface. At its base, there is a layer of anhydrite with nests and interlayers of gypsum and gypsum clays up to 14 m thick (Vodolazskaya et al. 2015). North of the Verkh-Yazva village, the outcrops of this layer form a chain of karst sinkholes up to 60 m in diameter. Some of them are filled with water (Fig. 3.15).

Intensive dissolution of salts within the anticlines led to the destruction of thick strata of rock and potassium-magnesium salts. At the same time, extensive subsidence troughs were formed along the anticlines. Thick strata of pebbles and loose sandy-clayey sediments indicate that riverine and fluvioglacial flows took part in the formation and filling of the troughs.

As a result of the intensive development of karst and other exogenous processes, the tectonic structure was not directly reflected here, as in other regions of the Cis-Urals, but a reverse reflection in the relief of the earth's surface. Small mineralization of lake and river waters indicates a decrease in the intensity of karst in the modern era.

In the area of the Novaya Zyryanka village, according to research carried out by the Mining Institute of the Ural Branch of the Russian Academy of Sciences (1999), JSC "Geokarta" (1999) and JSC "Galurgiya" (2000), as a result of karst and suffusion processes, 50 dips and 14 trench-like

Fig. 3.15 Karst lakes north of the Verkh-Yazva village (*photo by* N. Maksimovich)

Fig. 3.16 Funnel and cracking in the Novaya Zyryanka village, 1999 (Maksimovich and Pyankov 2012c)

depressions of the earth's surface were formed with a total length of about 1500 m (Figs. 3.16 and 3.17). Perhaps these phenomena are associated with the development of the VDPMS.

No caves in sulfate sediments of the region have been found so far.

Existing specially protected natural areas. Within the Solikamsky karst region, there is one protected area of regional significance associated with the manifestation of karst—the **protected landscape** of Nizhnevishersk with an area of 44,685 ha. It is located in the Krasnovishersk district, it was proposed for protection in 1994. The geomorphological appearance of the SPNA is closely related to the manifestation of the ancient salt and sulfate karst—the formation of the largest subsidence troughs and Lake Nyukhti.

Economic activity and environmental pressure. *Impact of the development of the Verkhnekamskoye potassium-magnesium salts deposit on the activation of sulfate karst.* VDPMS is one of the world's largest deposits of potash salts, where sylvinite, carnallite rock, and brines are mined. The geological reserves of the deposit are estimated at 96 billion tons for carnallite rock, 113 billion tons for sylvinites, and 4.7 trillion tons for rock salt. The salt layer of the deposit, which has the shape of a lens, with an area of about 8.2 thousand

Fig. 3.17 Linear subsidence zone on the northwestern outskirts of the Novaya Zyryanka village, 2007 (*photo by* N. Maksimovich)

a *b*

Fig. 3.18 Dips formed as a result of flooding of the mines of the Vernekamskoye potassium-magnesium salt deposit. **a**—Failure on the territory of BKPRU-1 on August 22, 2007 (*photo by* A. Starkov); **b**—a dip in the forest, July 27, 1986 (*photo by* N. Maksimovich)

km^2, is traced in the meridional direction for 206 km, in the latitudinal direction—up to 56 km (Kudryashov 2013). In the supra-salt layer, as mentioned above, sulfate rocks are widespread.

A sharp change in the hydrogeological and geochemical regimes during the construction and operation of potash mines could cause the dissolution of gypsum and then salts and the subsequent breakthrough of groundwater (Osipov et al. 2014a, b). In some cases, this leads to the flooding of mines (Baryakh et al. 2013, 2016) and the formation of sinkholes (Fig. 3.18).

Technogenic karst processes can develop in sulfate rocks to some degree.

Impact of waste from the potash industry. During the development of the deposit, significant volumes of enclosing rocks and potash production wastes of different phase compositions are stored on the surface. Up to 70% of all mined ore goes to sludge storages, salt dumps, and brine ponds. Solid halite wastes go to the salt dumps, and their content of sodium chloride is more than 90%. Clay material, the substance of the insoluble residue in the form of clay-salt sludge, and excess brines enter the sludge storage. Clay-salt sludge consists of water-soluble salts and insoluble clay sediment. Atmospheric precipitation and brines are filtered through salt dumps and sludge storages and form technogenic sodium chloride filtration effluents with salinity up to 300 g/dm^3 and neutral pH. Filtration effluents from dumps and sludge storages enter groundwater, determining the transformation of the chemical composition of the near-surface hydrosphere over a large area: the change from the hydrocarbonate-calcium type of surface and groundwater to chloride-sodium (Fig. 3.19; Maksimovich and Khayrulina 2012a; Khayrulina and Maksimovich 2018).

Fig. 3.19 Salt dump and brine tank on the territory of VDPMS (*photo by* N. Maksimovich)

The influence of waste from the potash industry leads to a physicochemical transformation of the void space of rocks, which can cause a kind of karst process.

Chloride-sodium groundwater has a greater dissolving capacity in relation to gypsum (see Chap. 1), which leads to an increase in porosity and permeability of fractured water-bearing rocks, and a change in the rate of salt migration (Liu 2012).

Modeling the change in the filtration parameters of the gypsum rocks during the filtration of sodium chloride brines in the area of one of the dumps, carried out by Yu. Liu and A. V. Lekhov, made it possible to assess the possibility of the development of technogenic sulfate karst (Fig. 3.20). The gypsum zone is a direct consequence of the degradation of the salt deposit due to dissolution by groundwater. Rock

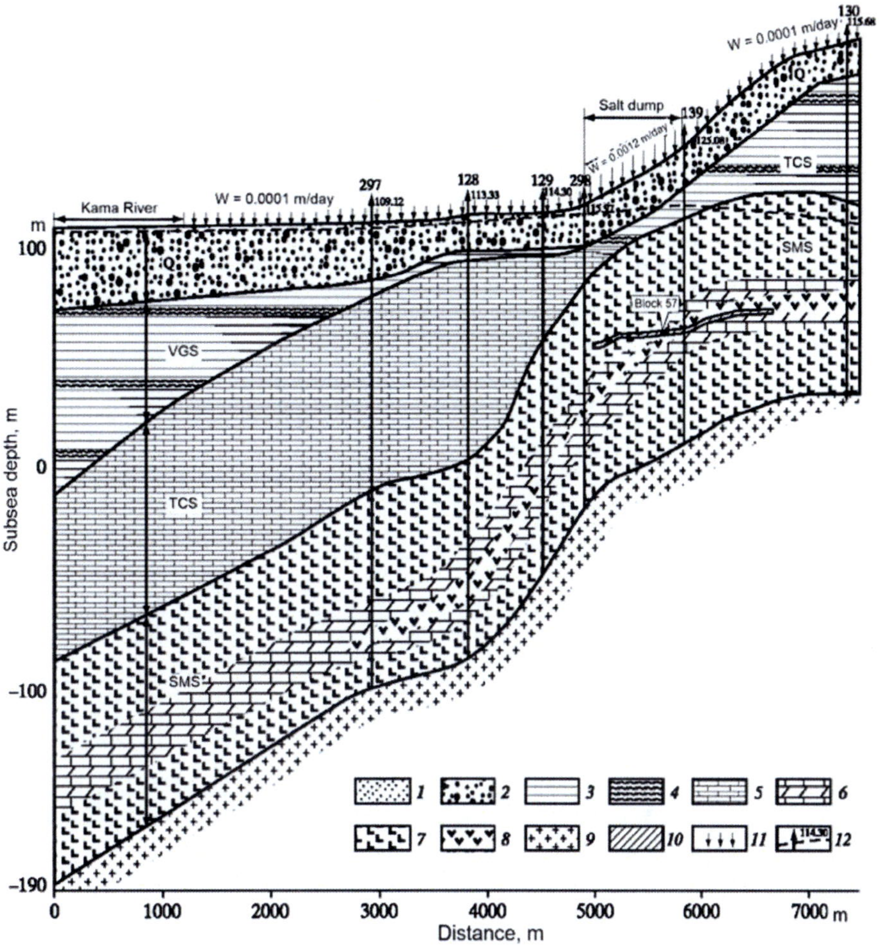

Fig. 3.20 Hydrogeological section in the area of the salt dump (Liu and Lekhov 2012). 1—Sand; 2—pebbles and gravel; 3—mudstone; 4—siltstone; 5—limestone; 6—marl; 7—marly clay; 8—gypsum; 9—rock salt; 10—test layer; 11—infiltration; 12—well and groundwater level

salt almost always contains anhydrite in the amount of 0.5–3.5 wt.% (on average 1.7%); upon joint dissolution, calcium sulfate migrates with brines and precipitates when the thermodynamic conditions change and the salinity of the solution changes.

On the territory of the VDPMS, in the cracks of rocks, zones of secondary minerals are distinguished: hydroxides of iron and manganese, calcite, halite, and gypsum. In the upper part, there is a layer of almost pure gypsum 1–2 m thick, below there are also interlayers or lenses of gypsum (Liu and Lekhov 2012).

Long-term filtration of sodium chloride brines in rocks with a soluble fracture filler (gypsum, calcite) leads to a change in their porosity and permeability. According to the modeling data, the zone of gypsum rocks 4–8 m thick above the gypsum mirror is completely deprived of gypsum. A narrow zone of a strong increase in porosity and permeability from 0.03 to 50 m/day can form on the surface of the gypsum layer. The influence of the variable density of groundwater due to the influx of brine leads to the lowering of the lower boundary of the brines filtering from above, and to the intensification of the process of dissolution of gypsum on the roof of the gypsum mirror by 10–20%.

The zone of gypsum rocks located above the gypsum mirror and the roof of the gypsum mirror itself in the vicinity of the salt dump undergo the greatest transformations. This leads to an acceleration of the migration of saline waters in the direction of the main drain of the Kama River (Liu 2012).

Influence of interstratal flows. The activation of sulfate karst is possible due to interstratal flows into the freshwater zone and outflows to the surface of highly mineralized groundwater from the supra-salt strata along the shafts of numerous wells (structural, hydrogeological, exploration, etc.). Throughout the history of exploration and development of the field, the number of such wells has reached tens of thousands, and documentation on many of them, including information on their location, has been lost.

The assessment of the possibility of such flows and outflows was carried out on the basis of the results of the analysis of hydrodynamic and hydrochemical conditions and sampling of springs in the region of the Nizhnezyryansk

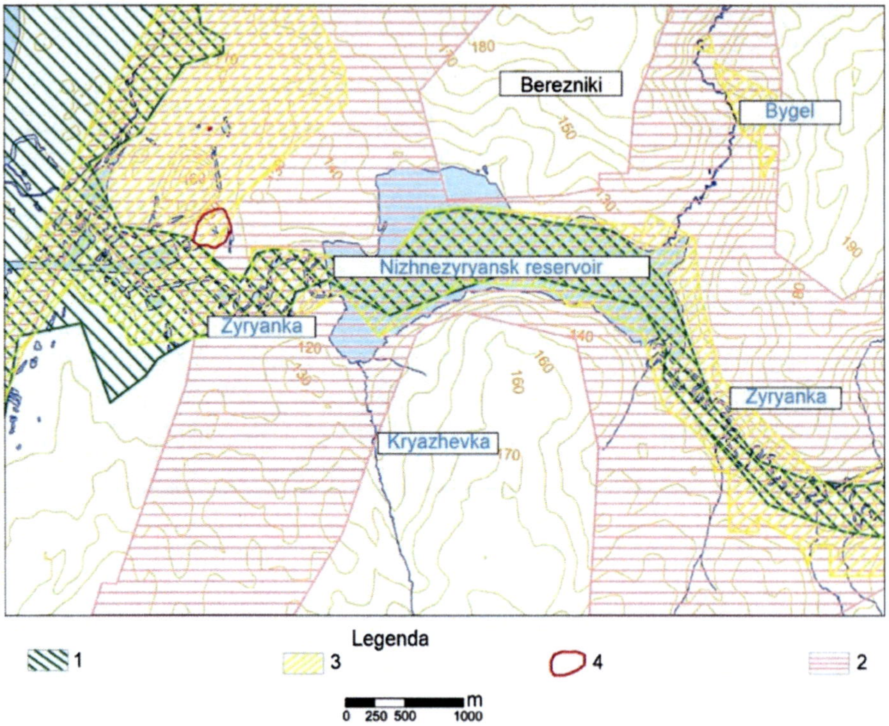

Fig. 3.21 Zones of possible overflows and outflows of groundwater in the area of Berezniki town (Maksimovich and Pervova 2012b). 1—Zone 1. Groundwater flows from the TCS to the surface are possible; 2—Zone 2. Groundwater overflows from the SMS to the surface are possible; 3—Zone 3. Overflows of groundwater from SMS to TCS are possible; 4—Failure

reservoir, located in the central part of the VDPMS (Fig. 3.21; Maksimovich and Pervova 2012b; Royeret al. 2017). It was created in the 50s of the last century on the Zyryanka River to supply water to the surrounding enterprises and regulate the river flow.

Sampling of springs in the coastal zone of the reservoir showed that, according to their chemical composition, there are three groups of springs. The waters of the first group—fresh with mineralization up to 0.3 g/dm^3—are discharged from the terrigenous-carbonate strata; the second group—brackish with mineralization up to 3 g/dm^3—are unloaded from the Upper Solikamsky aquifer; the third, most numerous group—saline with a salinity of up to 30 g/dm^3—are unloaded from the Lower Solikamsky aquifer.

On the left bank of the Nizhnezyryansk reservoir, 100 m from the springs, according to archival data, there was a well (Fig. 3.22). Here, the streams flowing from the springs do not have pronounced runoff channels, but the areal discharge of groundwater is observed (Maksimovich and Pervova 2012b), i.e., crossflows through the well began relatively recently.

Comparison of the data of bathymetric surveys in 2007 and 2020 showed that a relief deformation zone with a maximum depth of 4.5 m appeared at the bottom of the reservoir during this time.

The largest number of springs in the third group are clearly of technogenic origin and are located in the tail section of the reservoir (Fig. 3.23). Old wells are also located here. Attention is drawn to the high content of sulfates up to 0.7 g/dm^3 and calcium up to 0.5 g/dm^3 in comparison with other springs, which is associated with the dissolution of sulfate rocks of the supra-salt strata. It should also be noted that over the past 12 years, the mineralization of springs associated with the overflow of water from deep horizons has more than doubled, inclusively due to sulfates and calcium. Here, areas with halophytic vegetation have formed, which include such species as saline quinoa (Spergularia salina), rejected quinoa (Atriplex patens), spreading ratweed (Puccinellia distans), seaside triacle (Triglochin maritimum), Tatar lettuce (Lactuica), marsh plantain (Plantago uliginosa), etc.

In the area of the well, a zone is localized where the total subsidence of the earth's surface reached 800 mm.

The data on the chemical composition of the waters allow us to speak about the manifestation of sulfate and salt karst here (Maksimovich and Pervova 2012b).

Hydraulic engineering. Resolution of the Council of People's Commissars of the USSR and the Central Committee of the All-Union Communist Party of Bolsheviks dated August 10, 1937, No. 1339 "On the construction of the Kuibyshev hydroelectric complex on the Volga River and hydroelectric facilities on the Kama River" provided for the exploration and construction of the Verkhnekamsk (Solikamsk) hydroelectric power station on the Kama River and near the Solikamsk town and the creation of a reservoir for regulation of the flow of the upper Kama, the Vishera and

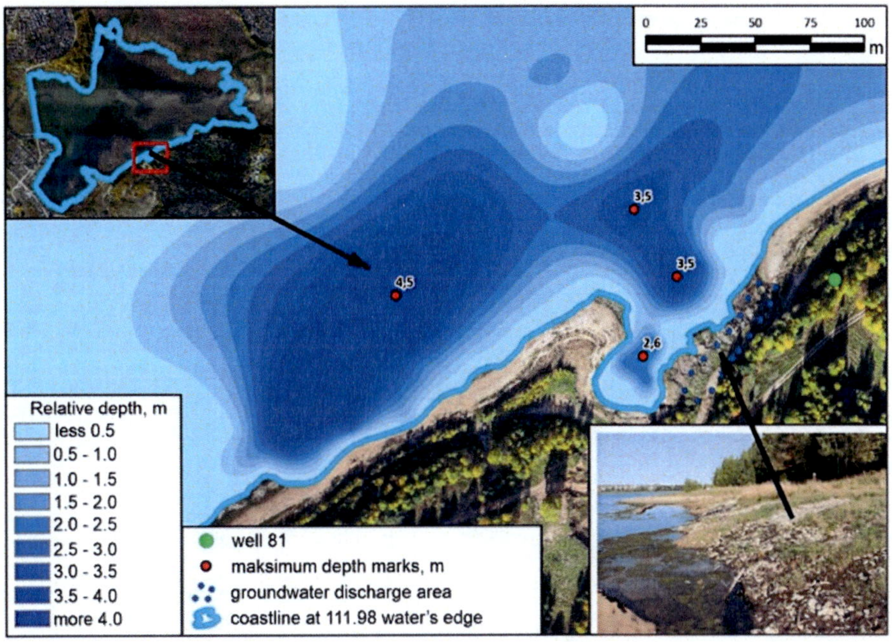

Fig. 3.22 Formation of deformations at the bottom of the Nizhnezyryansk reservoir in the area of outflows of mineralized groundwater from deep horizons

Fig. 3.23 Zone of intensive discharge of saline groundwater with a high content of sulfates in the tail section of the Nizhnezyryansk reservoir (*photo by* N. Maksimovich)

the Kolva Rivers, taking into account the possibility of future discharge of the Pechora River to the Kama and the Volga basins. To solve the Kamo-Pechora problem, three hydroelectric complexes were to be built—Solikamsky, Verkhne-Vychegodsky, and Verkhne-Pechorsky, forming a reservoir with a volume of 140 billion m^3 and an area of 16,000 km^2 (Lavinskaya and Khlevnyuk 2008).

On September 11, 1940, the Council of People's Commissars of the USSR issued a decree "On the organization of the Main Directorate of Hydraulic Engineering Construction of the NKVD of the USSR", which adopted a decision on the temporary conservation of the Solikamsk hydroelectric complex. As a result, it was never built, however, during the design, the conditions for the possibility of preserving highly

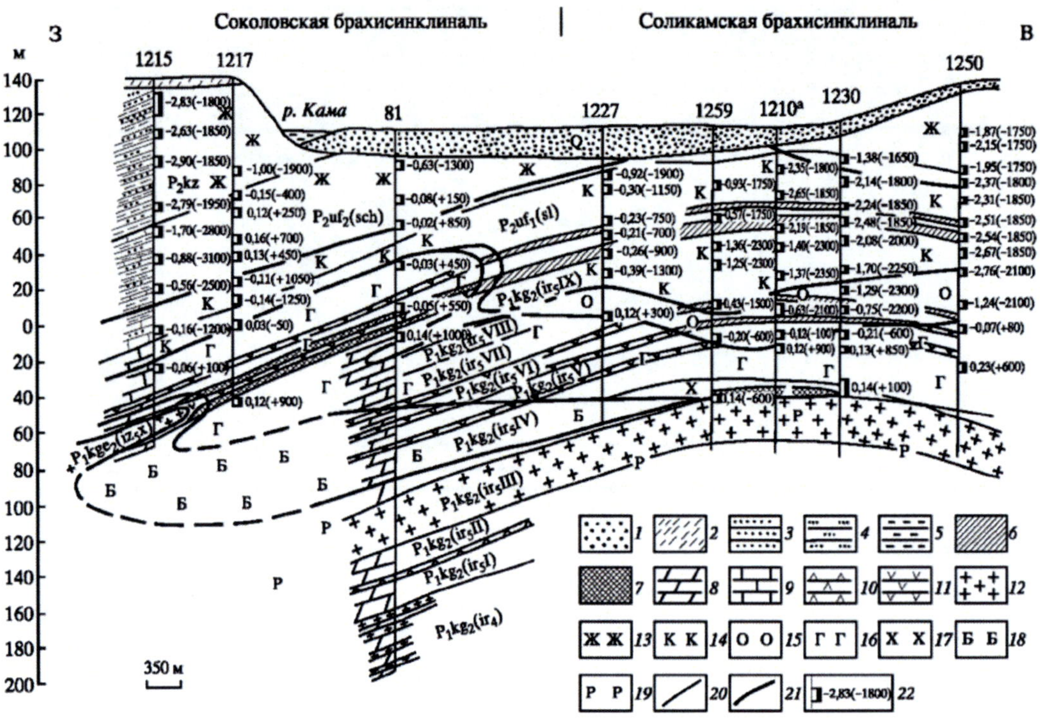

Fig. 3.24 Saturation degree and saturation of groundwater in relation to gypsum (Osipov et al. 2014a, b). 1–12—Lithological characteristics of rocks (1—sand, 2—loam, 3—sandstone, 4—siltstone, 5—mudstone, 6—brecciated, calcareous clay, 7—brecciated, dolomitic clay, 8—marl, 9—limestone, 10—anhydrite, 11—gypsum, 12—rock salt); 13–19—the composition of the mineral filling of cracks in rocks (13—iron and manganese hydroxides together with calcite, 14—calcite, 15—the absence of mineral filler in open cracks, 16—fibrous gypsum, 17—chalcedony together with fine-crystalline gypsum, 18—colorless halite, 19—pink halite together with chalcedony and fine-crystalline gypsum); 20–21—boundaries (20—lithological and stratigraphic, 21—interfaces between zones with different compositions of mineral filling of cracks); 22—the interval of the hydrogeological experiment (the first number is the degree of saturation according to the SOLMINEQ.88 program, the number in brackets is saturation, mg/dm^3; plus—water is oversaturated, minus—water is not saturated)

soluble rocks during filtration under the dam were studied in detail (Osipov et al. 2014a, b). The base of the projected section is composed of gypsum and saline rocks (Fig. 3.24).

The waters of the first aquifer are generally not saturated with respect to gypsum. In the second aquifer, groundwater is supersaturated in relation to gypsum. The groundwaters of both aquifers are not in equilibrium with halite. In the section, there are sharp hydrogeochemical boundaries between unsaturated and saturated groundwater in relation to the main hydrogenic minerals (calcite, dolomite, gypsum, halite). As world practice shows (James and Kirkpatric 1980; Milanović et al. 2019a, b; Maksimovich and Meshcheriakova 2017; 2018; Maksimovich et al. 2017; Maximovich and Meshcheryakova 2009a, b, c), in the presence of evaporites at the base of dams, serious anti-seepage measures are required, which leads to a significant increase in the cost of construction, and in some cases—to accidents and even disasters.

3.4 Polaznensky Region of Predominantly Sulfate and Carbonate-Sulfate Karst

The area occupies the central part of the Perm Region and has an area of 3102 km^2. From the west, it captures a part of the right bank of the Kama reservoir, from the south it is limited by the valley of the Chusovaya River, and from the north and east—by the basins of Dobryanka, Vilva, and Usolka. The northwestern border runs along the contact of the karst Solikamsky rocks with the Sheshminsky ones, the northern and eastern ones have complex outlines and are determined by the lithology of the rocks. Administratively, the region is located in the Chusovskoy, Dobryanka, Ilyinsk, Krasnokamsk, and Perm districts of the Perm Region. In the east, in the valleys of the right tributaries of the Vilva and Usolka Rivers, sulfate and carbonate rocks are replaced by terrigenous rocks (Fig. 3.25; Gorbunova et al. 1992).

Fig. 3.25 Polaznensky karst region

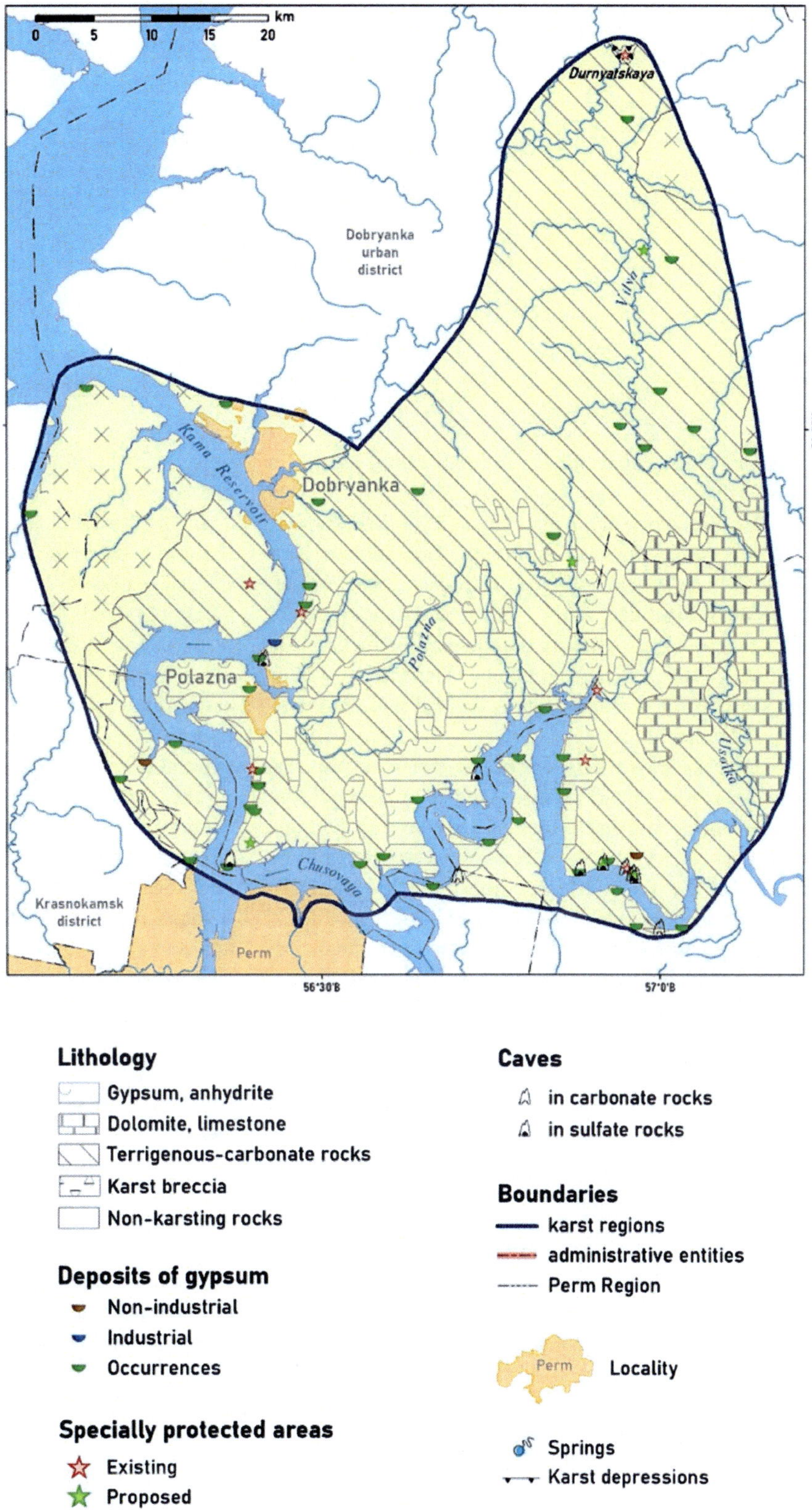

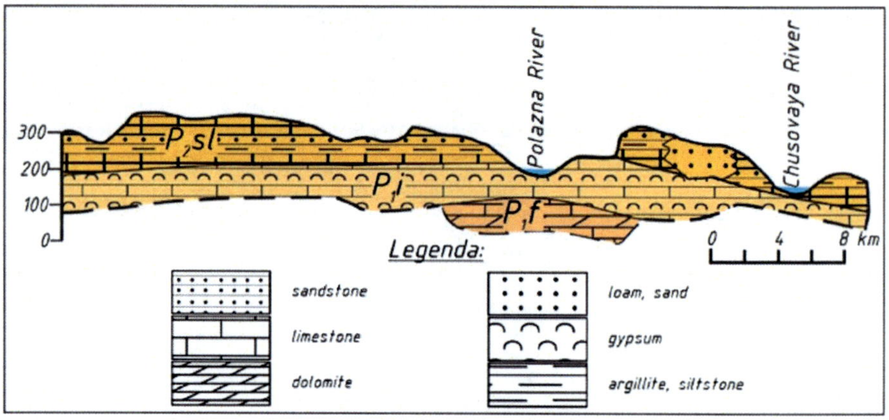

Fig. 3.26 Geological conditions for the development of karst in the Polazna region (Gorbunova et al. 1992)

Geological conditions for the development of karst. The area is confined to the Kosvinsko-Chusovskaya saddle, which occupies a part of the platform, and is traced eastward in the direction of the Cis-Ural trough. The saddle is complicated by the Mezhevskaya, Olkhovskaya, and Kamennolozhskaya swell-like zones. The zone of active water exchange and karst formation is composed of rocks of the Kungurian stage, Neogene, and Quaternary systems (Fig. 3.26). The Kungurian stage is divided into the Philippian and Irenian horizons.

Seven packs are distinguished in the Irenian horizon. The Upper Lunezhsky suite is composed of fine-grained gypsum, in places clayey with inclusions of large crystals of transparent gypsum. Anhydrite occurs in its lower part; the thickness of the layer increases from 40 m on the Sylva River to 65 m in the area of the Polazna village. The study of the section in the quarry of the Chumkassky gypsum deposit showed that in its lower part there is an indistinctly layered to massive anhydrite-gypsum rock (layer II) with an indistinct nodular structure (the size of nodules reaches 1–1.5 cm) (Kalinina 2015).

Anhydrite is noted in the form of fan-shaped and sheaf-like aggregates of long prismatic grains. Towards the top of the layer, the content of clay-carbonate material increases, and the texture of sulfates changes to nodular (layer III). A similar section is observed in the Shubinsky gypsum quarry in the Irensky karst region (Fig. 3.27).

The underlying carbonate Tyuy suite has a thickness of 1–6 m. The next—Demidkovskaya—suite, consisting of gypsum and anhydrite with interlayers of clayey dolomite and gypsum, has a thickness of 17 (the Polazna village)—47 m (the Golubyata village). It is underlain by the Yelkinskaya suite of clayey and oolitic dolomites, in places turning into dolomitized limestones with clay interlayers up to 15 m thick. The Shalashninskaya suite is represented by gypsum and anhydrite with clay and dolomite veins 15–33 m thick. increases from west to east from 2 to 15 m. At the base of the section, there is a Ledyano-Peshcherskaya suite of gypsum and massive anhydrite with dolomite interlayers up to 66 m thick.

In the western part of the region, the deposits of the Kungurian stage are overlain by rocks of the Solikamsky suite of the Lower Permian, which are represented by platy limestones, marls, dolomites, sandstones, gypsum, and in the eastern, in the Vilva River basin, by limestones, sandstones, conglomerates, mudstones, dolomites. Within the Chumkassky gypsum deposit, the Solikamsk deposits overlap the Lunezhsky sulfates with a stratigraphic break and contain gypsum (Fig. 3.28, layer II–V; Kalinina 2015).

The Sheshminskaya formation of the Lower Permian within the region is found in the form of small areas and is represented by mudstones, siltstones with interlayers of sandstone and limestones. The rocks are plastered. Gypsum is found in the form of lenses, interlayers, individual crystals, and also in a dispersed state.

Karst breccias occupy 750 km^2 (24% of the area of the region) and consist mainly of fragments of limestone, dolomite, marls with calcite or lime-clay cement up to 100 m thick and more, often represented by loose loams with fragments of carbonate rocks. The karst breccias contain remnants of interlayers and units of carbonate rocks of the Irenian horizon, as well as sand and gravel from terrace deposits (Lavrova 2005).

Basically, limestone and dolomite breccias were formed as a result of leaching of gypsum and anhydrite and the subsequent collapse of carbonate rocks into karst cavities.

Hydrogeological conditions. The zone of active water exchange, within which karst occurs, is confined to the Quaternary, Sheshminsky, Solikamsky, and Irensky aquifers and complexes. The Quaternary and Sheshminsky aquifers are developed locally. The Solikamsky complex is more widespread. Fractured limestones, sandstones, marls occurring on unkarst gypsum, and anhydrite of the Irenian horizon are aquifers. Where the Solikamsky deposits lie above the erosional cuts, and the underlying gypsum and anhydrite are strongly karst, anhydrous areas are noted. Continuous water

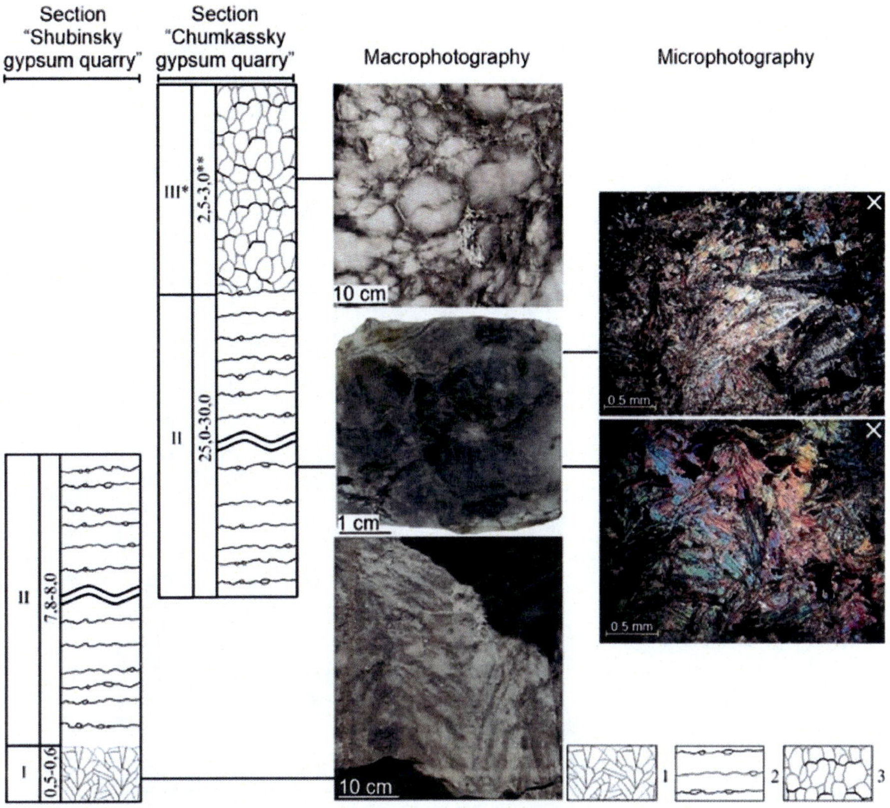

Fig. 3.27 Lithological section of the Lunezhsky suite (Kalinina 2015). Sulfates: 1—with a relic chevron structure; 2—massive, 3—nodular

cut is noted in the backwater zone of the Kama reservoir and river valleys (Gorbunova et al. 1992). The waters are characterized by a lack of pressure. The unloading is mostly dispersed, often outlets of groundwater swamp vast areas of the valleys.

A significant part of the shores of the Kama reservoir within the region is composed of karst rocks—gypsum, anhydrite, and limestone (Fig. 3.29).

The Polaznensky karst region almost completely includes the Chusovsky and Kamsky reaches of the reservoir for about 240 km (Actual problems… 2004). The total length of the coastline of the reservoirs of the Perm Region is 3.5 thousand km, of which 25% passes through the territory of karst regions (Maksimovich and Meshcheriakova 2009a, b, c; Maksimovich et al. 2019).

The construction of the Kama reservoir in 1954 intensified karst processes. Significant changes in the conditions of karst development took place in the karst areas (Pecherkin and Zakoptelov 1982; Pecherkin 1969; Maksimovich and Bolotov 1982). After its construction, the water level in most of the region increased by 20–22 m, as a result of which the groundwater level rose significantly, especially in the coastal zone. Filling in the spring to the design mark and level drawdown by 6–7 m leads to a change in the hydrodynamic zoning of karst waters—a decrease in the size of the vertical descending circulation zone and an increase in the size of the horizontal circulation zone, as well as periodic inflow of low-mineralized waters into the massif.

A significant part of the section of sulfate rocks was in conditions of complete watering. The steep gypsum banks in the flooded zone undergo dissolution under the influence of river waters. The processes of dissolution of sulfate rocks are especially intensive in the zone of seasonal drawdown of the reservoir. Here, loose filler of karst cavities is actively eroded and carried out. Dissolution of karst rocks under conditions of their constant or periodic watering, and suffusion of the filling of cavities activate karst processes. Renewal of old and emergence of new karst forms takes place. Funnels, caves, and wave-breaking niches are formed, and the phenomenon of slope subsidence is widespread (Matarzin 2003; Pecherkin 1969). In shattered sulfate rocks, the corrosive action of water transforms networks of karst cracks into labyrinth caves. Such caves, as well as niches (Fig. 3.30a), rapidly evolve, breaking down as a result of the collapse of the vault (Fig. 3.30b).

The collapse of rocks is facilitated by the presence of units in them in the form of blocks. When the coast is destroyed, columnar remnants are separated from it (Fig. 3.31). Blocks of rocks, caught in water, quickly dissolve (Fig. 3.32). According to observational data, karst shores in the first years

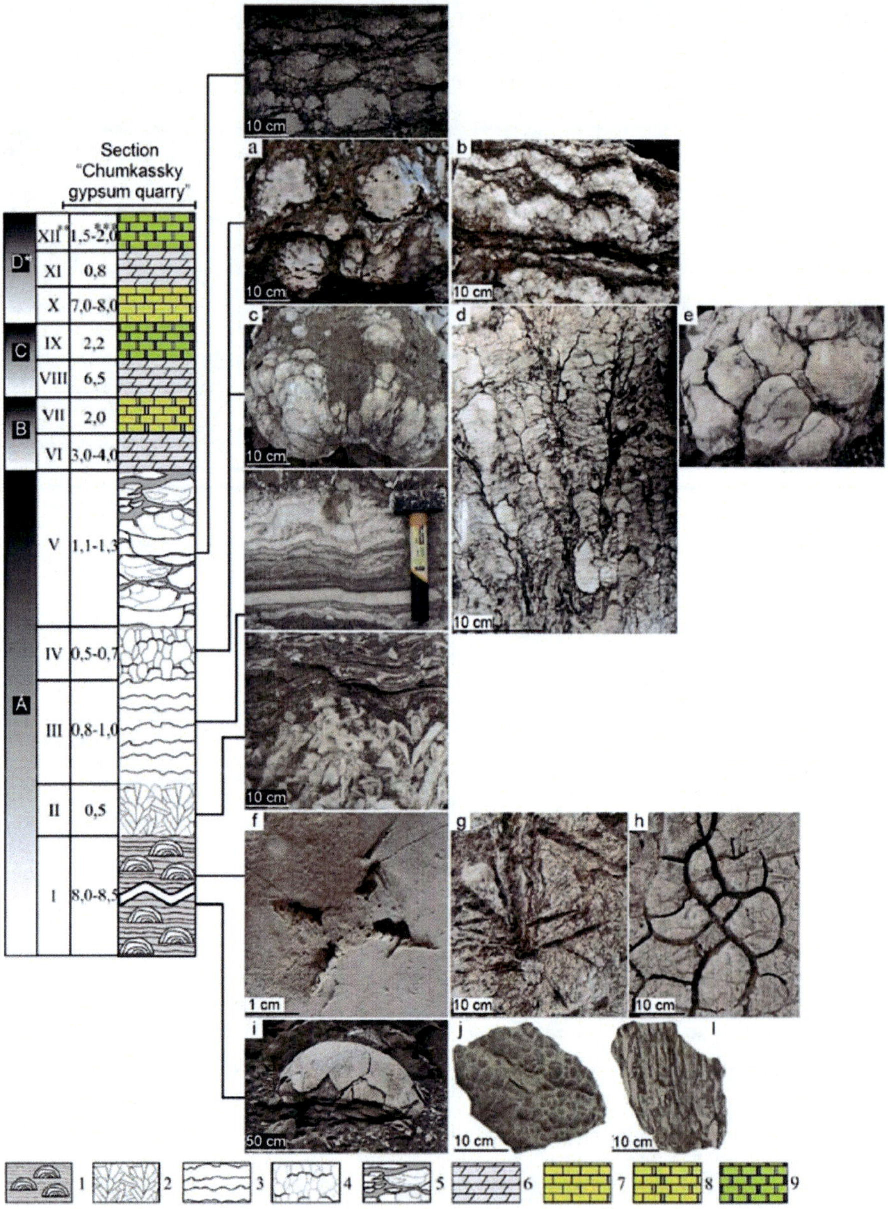

Fig. 3.28 Lithological section of the Solikamsky formation at the Chumkassk gypsum quarry (Kalinina 2015). 1—Thin-layered dolomite with stromatolite edifices; 2–5—textural and structural types of sulfates (2—with a relic chevron structure; 3—indistinctly layered; 4—nodular; 5—lenticular-nodular); 6—marl; 7—limestone; 8—dolomite limestone; 9—dolomite. Designations: **a**—gypsum nodules; **b**—enterolitic folding; **c–e**—aggregates of cauliflower-type gypsum nodules (**c**, **d**—longitudinal cut; **e**—transverse); **e**—imprint of a skeletal rock salt crystal; **g**—prints of ice crystals; **h**—drying cracks; and **l**—stromatolite structures (and—hemispherical; **k**—wavy-layered stratal; **l**—columnar)

after the creation of the reservoir retreated at an average rate of 2–2.5 m per year (Pecherkin 1969).

Above the outcrops of sulfate rocks along the banks of the reservoir in the area of Dobryanka town, the content of sulfates is 10–60 mg/dm^3, increasing at the dam site in the area of the Kama Hydroelectric Power Plant (HPP) to 20–100 mg/dm^3, including due to the influx of waters from the Sylvensko-Chusovsky reach, which is located in the area of intensive development of sulfate karst. A narrow coastal strip near the Polazna village is distinguished by an abnormal sulfate content (Kitaev and Rochev 2008).

The karst waters of the Irenian deposits are fed by rain and melt waters, which are absorbed by rock cracks and craters. Karst waters in gypsum and anhydrite have salinity up to 3.5 g/dm^3, sulfate composition, and high hardness. In the backwater zone of the reservoir, mineralization can decrease to 1 g/dm^3. The carbonate packs in well-washed areas contain fresh (up to 0.5 g/dm^3) hydrocarbonate waters (Gorbunova et al. 1992).

The flow rate of springs varies within wide limits from 0.01 to 40 dm^3/s with characteristic values of 1–5 dm^3/s.

a *b*

c *d*

Fig. 3.29 Outcrops of gypsum on the banks of the Kama reservoir **a**—in the area of the Mokhovo village; **b**—Chusovskoy reach, in the area of the Sokolya Gora tract; **c**—the right bank of the Polaznensky bay; **d**—specially protected natural area Lunezhsky mountains (*photo by* N. Maksimovich)

a *b*

Fig. 3.30 Formation (**a**) and destruction (**b**) of wave-breaking niches, Chusovsky reach of the Kama reservoir in the area of the Zamenaya Gora tract (*photo by* N. Maksimovich)

a *b*

Fig. 3.31 Gypsum remnants on the bank of the Kama reservoir: **a**—in the area of the Mokhovo village; **b**—the right bank of the Polaznensky bay (*photo by* N. Maksimovich)

a *b*

Fig. 3.32 Collapse and dissolution of gypsum on the bank of the Kama reservoir in the area of the Polazna village (*photo by* N. Maksimovich)

All types of discharge of karst waters are observed in the region: spring, subaqueous at the bottom of lakes and rivers, underground—in other aquifers. Karst waters feed rivers where carbonate layers have been exposed by erosion (Gorbunova and Maximovich 1981).

Groundwater belongs to sulfate-hydrocarbonate-calcium-sodium, hydrocarbonate-calcium-chloride-sodium hydrochemical facies with salinity from 0.22 to 1.76 g/dm^3. Changes in the chemical composition and mineralization are due to the influence of the Kama reservoir, mixing with the waters of landslide-karst deposits and fractured-karst waters of the Irenian horizon. Sulfate–calcium or sulfate–calcium-hydrocarbonate waters are hydrochemically stable (Kilin and Minkevich 2018).

Wells in Quaternary and landslide-karst sediments reveal groundwater of different chemical compositions: hydrocarbonate-calcium-chloride-sodium with salinity up to 0.6 g/dm^3 and sulfate-hydrocarbonate-calcium-sodium with salinity up to 2.1 g/dm^3. The Irenian aquifer is sporadic. The discharge of highly mineralized waters is confined to the zones of tectonic faults. In places where there are no zones of tectonic disturbances, the aquifer has no water runoff in the reservoir, there is a process of infiltration into the coastal sediments, and the chemical composition corresponds to the water of the reservoir (Kilin and Minkevich 2018).

Hydrogeochemical sampling of groundwater in the coastal zone of the reservoir near the Polazna village was carried out by the authors in 2002–2011 (Meshcheriakova

2014). Wells with a depth of 40–80 m are located at a distance of 50–400 m from the reservoir. The water level in them practically coincides with the level in the reservoir (correlation coefficient 0.99). This indicates their direct hydraulic connection, which is due to the intensive karsting of the rocks of the coastal zone.

According to the observations, it was established that the groundwater has a sulfate-hydrocarbonate-calcium composition. Water salinity varies from 1.8 to 4.6 g/dm^3. The hydrogen index has a value of 6.4–7.9. The content of macrocomponents changes within the following limits (mg/dm^3): SO_4 612–1790; HCO_3 171–1281; Cl 8.5–516; Ca 361–782; Mg 9.1–273; $Na + K$ 5.1–685. Calcium is predominant among the cations, and sulfates are the predominant anions.

Based on the results of testing, the dissolving capacity of water in relation to gypsum was assessed using the methods of Skilman-Mac Donald-Stiff and V. P. Zverev (Yarkeeva 1999; Zverev 1967), which showed that the degree of saturation of groundwater with calcium sulfate is very high—80–100%. The temporal and spatial patterns of the distribution of this indicator are not visible. This is due to the uneven distribution of atmospheric precipitation throughout the year, daily and seasonal fluctuations in the reservoir level, leading to a change in the direction of water movement both into the karst massif and out of it.

According to Malakhova et al. (2004), in this area, the deficit in the saturation of groundwater with calcium sulfate of the top water type was 1.8 g/dm^3, groundwater—1.7 g/dm^3, groundwater confined to the terrigenous-carbonate fissured thickness of Solikamsky rocks—1.7–2.0 g/dm^3, fissure-karst waters—0.0032 g/dm^3.

Karst phenomena. The gypsum and anhydrite of the Irenian formation of the Kungurian stage are most karsted (Gorbunova 1977). Karst is mainly sulfate and sulfate-carbonate.

According to the degree of exposure of karst rocks, the conditions of occurrence, and the nature of the overlying sediments, in accordance with the peculiarities of the circulation of fissure-karst waters within the region, the following areas are distinguished:

- covered karst (in the upper part of the bedrock slope of the Kama River valley), caused by the activity of fissure-karst waters in the zones of vertical descending, transitional and horizontal circulation;
- covered karst (middle and lower parts of the bedrock slope of the Kama River valley); in this area, there are small areas of almost bare karst;
- overlapped karst (on the third above-floodplain terrace of the Kama River valley), caused by the activity of fissure-karst waters in the zone of horizontal circulation, turning into a sub-valley one (Malakhova et al. 2004).

A large number of karst forms (40–50) per km^2 are noted on the northeastern limb of the Polaznenskiy uplift and on the adjacent southeastern limb of the Mezhevskaya swell-like zone. The high karst content of these areas is due to both their tectonic structure and favorable hydrogeological conditions. This site is located in the zone of the Kama and the Polazna Rivers, which are hydraulically interconnected (Pecherkin 1986).

In the section of the trough separating the Pionerskoe and Polaznenskoe uplifts, surface karst forms are practically absent. The Polaznenskoe uplift is complicated by regional and local tectonic faults and cracks. Here, surface and underground karst forms are confined mainly to zones of thickening and intersection of tectonic cracks.

On the platform, vaults are the most karsted. Karst usually occurs when the thickness of the overburden is no more than 50–75 m (Gorbunova et al. 1992). Surface karst forms are represented by numerous sinkholes, karren, kettles, karst lakes, depressions, dry valleys, karst ravines, and dolines. Underground ones—in cavities and caves.

Most of the cavities are located in the zone of horizontal circulation of karst waters, therefore, in the areas of the near-watershed and watershed areas, the relative depths at which the cavities are located are 50–60 m. Only single forms are found at depths of 30–40 m. The average vertical amplitude of the cavities is 3–3.5 m, but some reach 8–12 m. Most of the gypsum cavities are filled with sandy-argillaceous material, in which vertical zoning is observed: the cavities at the top are usually filled with fine-grained sand with an admixture of argillaceous material, and at the bottom—with medium-grained and coarse-grained gravel. The activation of the karst-suffusion process contributes not only to the formation of new cavities but also to the release of old ones from aggregate and an increase in their volume due to the dissolution of the rock.

Another feature of the Polaznenskiy-type arched uplifts is that they are broken by cracks oriented along the strike of the structure. Significant karsting of valleys and ravines is also associated with cracks of the onboard resistance (Pecherkin 1969). Their characteristic feature is their openness and strikes parallel to the slope. They are channels through which surface waters are filtered through loose sediments into the depths, therefore, chains of karst sinkholes are often formed here, and karst develops intensively in the depths of the massif.

The massif of the Polaznensky region is characterized by a dissonant remnant relief, the amplitude of which reaches 35–40 m (Pecherkina 1983). Negative karst forms (inter-outlier saddles) were formed as a result of leaching by groundwater of gypsum and anhydrite from the Irenian horizon of the Kungurian stage in zones of increased tectonic fracturing, which was accompanied by subsidence and crushing of the limestones and marls of the Solikamsky horizon of the Ufimian stage that overlap them. Above the outliers, the Solikamsky rocks have been preserved in their natural occurrence. Within the interstation saddles, as a rule, they are completely destroyed (Maksimovich and Bolotov 1982).

According to the observations of the Mining Institute of the Ural Branch of the Russian Academy of Sciences, at the beginning of 2020, 79 sinkholes were recorded in the area. Most of the dips were found in the Polaznensky karst area, located on the left bank of the Kama reservoir in the vicinity of the Polazna village, and in the Shalashninsky karst region, which occupies the right-bank part of the Chusovsky bay of the Kama reservoir. The average dimensions of the dips are up to 3 m in diameter and up to 5 m in depth, of which 7 dips with a maximum size of 10 m in diameter and 6–10 m in depth. The largest sinkhole is 25 m in diameter and 15 m deep, occurred in 1951, and 2 km southeast of the Ivanovka village. The sinkhole, recorded in 1991 in the Gorodishche village, had a depth of 27 m, dimensions in plan 2.8 by 1.7 m (Kilin and Minkevich 2006).

In 1975, on the road leading to the crossing over the Chusovaya River, on the left bank of the river, a dip with a diameter of about 3 m was recorded.

In the Polazna village, on August 5, 2017, there was a collapse at the exit from Trukhin Street to Sportivny Lane, at the edge of the road (Fig. 3.33a). The shape is oval in plan, diameter is 9.2 by 7.8 m; in section, it is bowl-shaped, and the depth at the northern wall is 2 m.

On August 6, 2019, in the Polazna village on 20, Kamskaya Street, there was a collapse in the garden of a private residential building. The dimensions of the hole at the time of the survey were 3.5 by 4.1 m and the depth—9 m. In 1 month, the diameter of the hole doubled, after which it was filled up.

On May 15, 2020, a sinkhole formed in Tenisty Lane in the northern part of the village in the Mokhovoe-4 district, on the slope of the left bank of the river. The climbers are 0.8 km southwest of the estuary. The shape in the plan is round with a diameter of 2 m, and in the section—pear-shaped and the depth is 5.9 m (Fig. 3.33b).

The largest number of sinkholes in the area was recorded in the period from 1960 to 1980. An increase in the number of sinkholes during this period was associated with a rise in the level of surface and groundwater after the construction of the Kama HPP in 1954. The activation of karst processes occurs not only in the coastal zone but also extends over considerable distances towards the watersheds (Gorbunova et al. 1992; Lukin 1964; Pecherkin 1969).

On the territory of the region, Butyrina (1968) identified 12 karst areas with a total area of 4 thousand km^2.

The **Polaznensky site** mainly occupies the left-bank part of the Kama reservoir (Kamsky reach) in the vicinity of the Polazna village. K. G. Butyrina also referred to it as the right-bank part of the reservoir. Subsequent researchers identified this part as an independent area (Nazarov 1996).

On the site, there are karren, kettles, ravines, dry valleys, and lakes. Breccia-filled karst depressions have been identified (Gorbunova et al. 1992). The highest karst content—up to 1056–1072 sinkholes/km^2—is confined to the near-edge sections of the reservoir near the Besovo village, the Konstantinovka village, the Zuyata village, northwest of the Mokhovo village, etc. (Fig. 3.34). Their average density at the site is 60 forms/km^2.

The formation of cavities is associated with leaching, as a rule, of sulfate rocks by aggressive waters, the movement of which is facilitated by a system of sublatitudinal and submeridional strikes. The most dangerous are the sections of

a

b

Fig. 3.33 Collapses in the Polazna village: **a**—Sportivny lane, 2017; **b**—Tensity lane, 6, 2020 (*photo by* O. Kadebskaya)

Fig. 3.34 Sinkholes of the Polaznensky karst area: **a**, **b**—near the Besovo village; **c**—near the Penki village, **d**—northwest of the Mokhovo village, **e**—in the area of the Zuyata village (*photo by* N. Maksimovich)

the slopes of the Kama and Polazna River valleys, where the depth to the horizontal circulation zone is no more than 25–30 m. On the slope, the karst coefficient is 0.186, and the number of sinkholes per 1 km^2 reaches 270; on the watershed part, these values are, respectively, equal to 0.036–0.004 and 17–30 (Pecherkin 1969). According to drilling data, the coefficient of underground karst content ranges from 4 to 59%, on average for the strata—21%. The near-edge sections of the steep slope of the reservoir and the territories adjacent to the ravines are characterized by the highest karst coefficient, where the average density is from 152 to 272 forms/km^2. Karst sinkholes are smaller, fully or partially exposed on the sides of karst rocks, with the absence of lakes and sinkholes with a swampy bottom. Small cone-shaped funnels are located quite tightly and their sides often touch.

Part of the Polaznensky area, located on the right bank of the Kama reservoir, tectonically corresponds to the north-eastern end of the Krasnokamsko-Polaznensky swell (Fig. 3.35).

The highest density of karst forms is observed in areas where the gypsum-anhydrite layers of the Irenian horizon are most closely located at the day surface. The karst content of this part is generally high. The density of craters in the coastal zone is 300 forms/km^2. The average density of karst and suffusion-karst forms per 1 km^2 is up to 30–50.

According to the degree of karst formation, the territory can be divided into three parts: the northern part with an average density of forms less than 1 form/km^2 (sinkholes and subsidence are rare here), the central part—up to 40–50 forms/km^2, and the southern part—from 1–5 to 15–20 forms/km^2, sometimes up to 150. The most typical sizes of craters are 8–15 m in diameter, the depth does not exceed 5–7 m (Nazarov 1996; 2004).

The **Lunezhsky site** stretches along the high left bank of the Kama reservoir from the Lunezhki village to the Zapolye village. In the south, the density of sinkholes is about 30 forms/km^2, decreasing in the north to 5–10 forms/km^2. With distance from the reservoir, the density of karst forms

a

b

Fig. 3.35 A high abrasion scarp formed in the Irenian gypsum near the Khokhlovka village: (**a**), and its destruction (**b**) under the influence of the Kama reservoir (*photo by* N. Maksimovich) sharply decreases and amounts to approximately 1 form/km^2 or less (Nazarov 2004).

The **Divyinsko-Kamennolozhsky site** is located between the Chusovaya and Polazna Rivers.

At the northern end of the Krasnokamsko-Polaznensky and Kamenolozhsky swells, ancient sulfate karst is very widespread. Modern karst develops in rocks that occur mainly above the local basis of erosion, as well as under river beds, i.e., in the zone of active circulation of groundwater (Butyrina 1968).

The site is distinguished by a small thickness of the cover stratum, a significant outcropping of sulfate rocks, and widespread karst breccias. A characteristic feature of the territory is the insular development of surface karst forms. The places of their formation are confined to the pillars of sulfate rocks, separated by ancient karst depressions filled with karst breccias. The density of craters in the coastal zone of the site (Chusovsky reach) rarely reaches 30 forms/km^2 and most often fluctuates within 5–10 forms/km^2. Approximately half the coastline is represented by high steep slopes of gypsum and anhydrite outcrops (Nazarov 2004).

The **Yarinsky site** is located within the northern immersion of the Kamennolozhsky swell in the interfluve of the tributaries of the Kama, the Kosva, and the Chusovaya Rivers. In the west, the site is bounded by the meridional section of the Dobryanka River, and, in the east, the border is conventionally drawn along the watershed of the Vilva and the Dobryanka Rivers. Thick clay and loamy overburden sediments (more than 50 m) impede the development of karst. Karst forms—mainly funnels—are confined to the arches of small uplifts and the valleys of the Yarina and the Talaya Rivers. There are also erosion-karst ravines and lakes with atmospheric nutrition. In the area of the Yarino station, their density does not exceed 0.5 per 1 km^2. The karst is ancient, as evidenced by landslide-karst formations up to 50 m thick. No modern sinkholes have been found at the site (Gorbunova et al. 1992). Sulfated karst underground waters affect the Dobryanka River. The content of sulfate ion in it increases from 8 mg/dm^3 in the upper reaches to 226 mg/dm^3 below the mouth of the Yarinka River (Butyrina 1974).

The **Shalashninsky site** is located on the right bank of the Chusovsky reach of the Kama reservoir at the confluence of the Vetlyana, the Mutnaya, the Talaya, the Shalashnaya, and the Belovskaya Rivers. The banks of the reservoir, composed of gypsum and anhydrite, undergo intensive processing and dissolution, forming characteristic forms (Fig. 3.36).

With increasing distance from the reservoir, sulfate karst becomes carbonate-sulfate. The density of karst forms in the coastal part of the site reaches 50–70 forms/km^2 (Nazarov 2004). Tectonically, the site is located within the arch and southeastern wing of the Kamennolozhsky swell (Gorbunova et al. 1992).

Fig. 3.36 The coast of the Chusovsky reach of the Kama reservoir, folded with gypsum, in the area of the dacha cooperative Beloevskoye (*photo by* N. Maksimovich)

Within the boundaries of the four karst fields identified here with an area of 13.5 km^2, 175 sinkholes, 10 ravines, 3 karst depressions, 12 lakes, disappearing streams and rivers, ascending springs with a sulfate composition, fresh sinkholes are known.

The **Kulikovsky site** is located along the right bank of the Chusovsky reach of the Kama reservoir from the Nizhnie Penki village to the Konets Gor village. The coast is a series of fractured gypsum rocks with typical karst forms, including caves (Fig. 3.37). The average density of sinkholes in the coastal part is 5–10 forms/km^2 (Gorbunova et al. 1992; Nazarov 2004).

The site of the Chusovsky Cape is located between the Chusovaya and the Kama Rivers up to their confluence. Here sulfate rocks are overlain by platy limestones, marls, and dolomites up to 50 m thick (Fig. 3.38).

In some areas along the coastal zone, sinkhole fields with a high density are observed—up to 100 forms/km^2. With distance from the river valleys, the intensity of the karst decreases. Most of the sinkholes have a diameter of 9–20 m at a depth of up to 6 m. There are also larger sinkholes with a diameter of up to 130 m. The activity of karst is indicated by ponors at the bottom of the sinkholes and subsidence (Gorbunova et al. 1992).

The development of karst occurs most actively at the contacts of rocks of various lithological compositions, in this case, carbonate and sulfate strata. Hydrocarbonate waters are replaced by sulfate ones. Due to the influence of fresh waters of the Kama reservoir, karst is activated, which reduces the degree of stability of the karst territory. The chemical composition of groundwater in the Irenian horizon in sulfate

Fig. 3.37 Outcrops of gypsum in the coastal zone of the reservoir on the Kulikovsky karst area (*photo by* N. Maksimovich)

rock massifs changes quite sharply both in time and along the section; at the same time, in the zone of constant water saturation, it remains quite stable. Along with the intensive development of karst, the metamorphization of groundwater occurs. Their mineralization and hardness increase and an increased content of nitrates is observed. In the karst basin of the Chusovsky Cape, swampy lakes with increased mineralization and high sulfate content are formed (Minkevich et al. 2019).

Karst sinkholes occur in the area. The Gorodishchensky depression, 27 m deep, was formed in 1991 in the zone of transitional circulation of groundwater. Initially, it was a shaft with a cross-section of 2.8–1.7 m (Tyurina et al. 1994). A depth of the sinkhole ended in a rounded grotto 13 by 8 m. In 2006, the depth of the sinkhole decreased to 7 m due to the collapse of the rock mass (Fig. 3.39). By 2018, the hole was completely covered with household waste (Kilin and Minkevich 2018).

The **Nizhnekosvinsky site** is located partly in the basin of the lower course of the Kosva River within the Kosvinsko-Chusovsky arch. In the Solikamsky horizon, which has a predominantly carbonate composition, there are interlayers of gypsum up to 6 m thick, as a result of leaching of which karst sinkholes are formed, some of which are conical and have a diameter of up to 75 m and a depth of up to 20 m. Terraces of the Kosva River are cut by erosion-karst ravines.

On the floodplain of the Pozhva River, there is the Durnyat karst depression with an area of about 2 km^2, where craters and 11 lakes are noted. According to the nutritional conditions and chemistry, the lake waters are divided into three groups.

The central group includes the deepest waste lakes (Beloe and Rogalek, 46 and 61 m deep, respectively) with a predominantly underground supply. They have sulfate and chloride compositions of water with salinity from 1.8 to 12.9 g/dm^3 (Fig. 3.40a; Table 3.2). The lakes of the western and eastern groups are in different stages of development: among them, there are recently emerged—Novy Proval, as well as those covered with raft—Bolshoye, Maloye, Savushkin Lozhok, Sidorovskoye (Fig. 3.40b, c). They are drainless and contain weakly mineralized water (Gorbunova 1985).

In all lakes, a decrease in the surface area and depth is observed due to sedimentation, waterlogging, and an increase in rafting. From the 60 s of the twentieth century to the beginning of the twenty-first century, the area of Lake Bolshoye decreased by about 1600 m^2, the depth—by 1 m. The area of White Lake decreased by 1100 m^2. The salinity of the lakes practically did not change (Minkevich and Oshchepkov 2004).

The discharge of waters from the Durnyatskaya depression occurs in the form of a stream flowing out of Lake Beloye, with a flow rate of about 500 l/s and a salinity of 5.5 g/dm^3 with a sulfate content of 1.5 g/dm^3 (Fig. 3.40 g). After the stream flows into the Pozhva River, its

Fig. 3.38 Cape Chusovskaya Strelka (confluence of the Kama and Chusovaya Rivers), composed of carbonate rocks overlapping gypsum, which were exposed here before the creation of the reservoir (*photo by* N. Maksimovich)

Fig. 3.39 Gorodishchensky karst sinkhole, 2006 (*photo by* Yu. Kilin)

mineralization increases 1.6 times. Subsequently, the salinity of the river also increases to 1.2 g/dm^3, which indicates an intensive subaqueous discharge of karst groundwater (Table 3.2).

The **Taborsky site** is located in the basin of the lower reaches of the Vilva River. It also includes the territories to the east of the Tikhaya village and the basin of the middle reaches of the Klyuchanka River. The site is located in the northern part of the Kosvinsko-Chusovskaya saddle. According to drilling data, among the sandstone-argillite-siltstone sediments of the upper Kungurian, sulfate rocks up to 20 m thick and lenses of rock salt up to 50 m are found here. Sulfate and salt-sulfate karst is of a covered type. On an area of 5.8 km^2, 253 karst forms were mapped (Fig. 3.41), mainly craters with an average density of 43, maximum—up to 450 forms/km^2 (Butyrina 1975).

According to local residents, in the twentieth century, 12 sinkholes and subsidence occurred near the Tabory village and the Boyanovo village within a radius of 2–2.5 km. In

Fig. 3.40 Water bodies of the Durnyatskaya karst depression. **a**—Flowing karst Lake Beloe; **b**—overgrown karst lake; **c**—drainless karst Lake Bolshoye; **d**—Istok stream flowing from Lake Beloe (*photo by* N. Maksimovich)

Table 3.2 Chemical composition of lakes and rivers of the Durnyat karst depression, September 17, 2013

Sampling location	pH	Content, mg/dm^3								
		HCO_3	SO_4	Cl	Ca	Na	K	Mg	Fe_{total}	Mineralization
Lake Beloe	7.24	273.0	1314.0	1805.0	641.0	1161.0	17.2	69.3	0.31	5157
Pozhva River, 20 m above the mouth of the Istok stream	8.02	251.6	43.4	34.6	75.9	26.9	<0.5	15.0	0.23	448
stream Istok	7.19	283.7	1522.6	1943.8	627.5	1053.0	<0.5	80.1	0.10	5511
Pozhva River, 200 m above the mouth of the Istok stream	7.73	247.8	115.8	125.4	111.3	81.4	<0.5	18.0	0.12	700
Pozhva River, Kosyata tract, bridge	7.84	251.6	281.7	330.7	156.9	201.6	0.6	24.9	0.09	1248
Pozhva River, Sofronyata tract, bridge	7.98	259.3	250.5	296.4	151.8	171.1	<0.5	24.7	0.09	1154

Fig. 3.41 Karst sinkholes north of the Tabory village (*photo by* N. Maksimovich)

1911, small earthquakes were noted here, which are possibly associated with karst sinkholes (Maksimovich 1964).

The **Golubyatskiy site** is located in the basin of the middle reaches of the Vilva River, where the Kosvinsko-Chusovskaya saddle is complicated by positive structures, within which gypsum and anhydrite come to the surface along the Kukhtym, the Peluch, and others rivers.

From karst forms, there are sinkholes, ravines, dry valleys (Kukhtym, Peluch, Sukhaya, and Mutnaya), karst depressions, springs, and numerous lakes. The diameter of the craters is 10–25 m (the maximum is up to 80 m) and the depth is 5–17 m. Outcrops of Irenian gypsum are noted on the sides and at the bottom of some craters (Butyrina 1975).

In the valley of the Vilva River, there are 7 known lakes with a diameter of 40–250 m and a depth of 4–12 m. The lakes are in different stages of development, they have predominantly sulfate water with a salinity of 2–3 g/dm^3 and more (Fig. 3.42).

The Usolka area is located in the basin of the Usolka River and the interfluvial spaces adjacent to it from the west and north. The southern wing of the Kosvinsko-Chusovskaya saddle passes here into the Yuryuzan-Sylvinskaya depression. Surface karst forms, mainly sinkholes, are found in groups or small fields. Near the Borisovo village, on an area of 1.7 km^2, there are 21 sinkholes with a diameter of 3 to 77 m and a depth of 1–2 m. Groups of sinkholes of two or four are confined to zones of increased fracturing of the Solikamsky deposits. Some are filled with water (Gorbunova et al. 1992). There is a swampy karst depression at the mouth of the Usolka River (Fig. 3.43).

The territory on the left bank of the Chusovaya River in the Golovanovo and Bannaya Gora districts of the Perm city, located partially on the IV terrace and slope (roughly horizontally with a mark of 160 m), not previously identified by K. A. Gorbunova, also belongs to the karst-hazardous territory (Utemova and Papirovaya 2015).

During surveys by "Lengiprotrans" in 1951, karst sinkholes were noted within the Bannaya Gora district for the transfer of the Gornozavodskaya railway from the flood zone. On the territory of the Perm Pulp and Paper Mill in the Golovanovo district, on the surface of the above-floodplain terrace III, a karst sinkhole measuring 6 by 6 m, 0.6 m deep and a buried sinkhole 20 m in diameter filled with peat and silt to a depth of 5.8 m were recorded. On the topographic

Fig. 3.42 Karst lake of complex geometric shape with a maximum diameter of 140 m in the north-east of the Vilva village (*photo by* N. Maksimovich)

Fig. 3.43 Karst depression in the estuarine part of the Usolka River valley (*photo by* N. Maksimovich)

plan of 1958, at the foot of the slope of terrace IV of the Chusovaya River, at elevations of 120–125 m, there were depressions in the relief of a rounded shape, up to 20 m in diameter, presumably of karst origin. Here, while drilling on the above-floodplain terrace IV at a depth of 42.5 m, the top of the anhydrite of the Irenian horizon of the Kungurian stage was uncovered. The exposed thickness of sulfate rocks was 20 m (Utemova and Papirova 2015).

Based on the data presented, this part of the territory of the Perm city is included in the Polaznensky karst region (Fig. 3.25).

Caves. Most of the caves of the Polaznensky karst region, located in the coastal zone of the Kama reservoir, are short-lived. The duration of their existence is determined by the rate of coastal processing (Pecherkin et al. 1965).

According to the results of studies carried out in 2018–2020, it was established that part of the previously described karst caves, such as Garmonovskaya, Zagoskinskaya, Kulikovskaya Novaya, Kulikovskaya Teplaya, Polazninskaya, Polazninskaya-2, Sinyavskaya, is completely destroyed, and another part is inaccessible due to the overlap of the entrances by the landslides of the coastal slopes (Malaya Divya). Also, no caves have been identified that were noted in the late 90s—early 2000s by observers who did not publish their materials—these are the Voronikhinskye and Chumkasskaya caves. Some underground karst occurrences have significantly changed in length and shape, for example, Kulikovskaya Ledyanaya.

The Malaya Divya Cave deserves a special mention, the total length of the passages of which, according to the survey materials of the Perm Speleoclub in 1984, was about 1000 m (Gorbunova et al. 1992). The cave had five entrances located in a rocky outcrop of gypsum and anhydrite, 12 m high, at the water edge of the Chusovaya River near the Malaya Divya tract. It was a labyrinth of teardrop-shaped passages 0.7–0.8 m wide and up to 1.5 m high. When cavers from Perm visited the area in 1991, it was found that the entrance area was blocked by a slope collapse, which is a mixture of terrigenous-sulfate rocks. In March 2019, the entrance to the cave was also closed.

The Kulikovskaya Cave (Kulikovskaya Ledyanaya) is located 100 m northwest of the outskirts of the Kulikovo village, on the I terrace above the floodplain of the right bank of the Chusovaya River.

At the time of research (May 2020), the cave had three entrances and was a single extended grotto. Previously, the cave consisted of two grottoes connected by an arched entrance 4 m wide and 6 m high. In 1961, there was a sinkhole over the small grotto, as a result of which an entrance to it through a sinkhole was formed.

According to I. Yu. Gerasimova's observations, in 2013, the upper entrance (through the well) was blocked by clay-rubble-block debris and reopened, presumably, in 2019 (Fig. 3.44).

The main entrance, 2.5 m high and 4.5 m wide, is located at the bottom of a sinkhole 8.5 m in diameter, at the base of the rocky outcrop of gypsum anhydrites (Fig. 3.45a).

Currently, an inclined descent leads into a grotto, elongated in a northeastern direction, in the western part of which there is a lake that goes under the walls of the cave, and, in the eastern part, there is a clay-rubble-block debris and two entrances, which previously represented a single well—the entrance into the Kulikovskaya Ledyanaya Cave (Bogoslovsky 1940) (Fig. 3.45 b).

In the western part of the grotto, near the main entrance, there is a low course filled with lake water and clay. According to the data of the semi-instrumental survey in 2020, the length was 73 m and the amplitude was 12 m.

According to archival data (Bogoslovsky 1940), the entrance to the nearby **Kulikovskaya Teplaya Cave** was previously located at a height of 5–6 m above the water's edge. Later, G. N. Panarina (1973) pointed out that the entrance to the cave was located at the water's edge and opened with an arched niche. At the time of research (May 2020), the entrance to the cave was not found.

The active development of the caves located near the coastal rock outcrops, the presence of a visually traced passage deep into the massif below the lake level in Kulikovskaya Ledyanaya, suggests the presence of a hydraulic connection between the lake and the reservoir, as well as the possible unification of the Teplaya and Ledyanaya Kulikovskie caves in the underwater part.

The **Beloy Gory Cave** is located in a rocky outcrop of carbonate and sulfate deposits, 1.4 km upstream from the Romakhino village. The entrance, 0.9 m high and 9.2 m wide, located at the water's edge, is a niche (Fig. 3.46a). It develops along a system of cracks with a predominant direction parallel to the river bank (Fig. 3.46b).

According to the data of the semi-instrumental topographic survey in 2020, the length of the cave was 40 m and the amplitude was 2 m (Fig. 3.47).

There are references to caves with the same name in this area (Maksimovich 1947; Pecherkin et al. 1965, 1981). When surveyed in 2020, they were not found. According to the description, only the cave surveyed by I. A. Pecherkin is similar to it, but there are some differences.

The **Konets Gor Cave** is located on the right bank of the Chusovaya River, 360 m upstream from the outskirts of the Konets Gor village.

Earlier (Pecherkin et al. 1980, 1981), the entrance was 6 m above the river's edge, and the cave consisted of two grottoes connected by an impassable channel. The accessible part was a winding, gradually tapering corridor with a length of 30 m.

In 2020, the entrance had a triangular shape (width 1.5 m at the base and height 0.7 m), was partitioned off by a block of bedrocks, and was at a height of 0.5 m from the water's edge (Fig. 3.48).

It is a crack of a skirting squelch 2.5 m wide and up to 3 m high. The crack gradually narrows, ending in a fall of clay-gravelly detrital material. There is a hydrodynamic relationship between the water levels in the cave and the reservoir. The total length of the cave is 12 m.

The **Sokoliy Kamen Caves** (1–6) are located within a 70 m section, 850 m downstream of the Chusovaya River from the outskirts of the Ust-Shalashnaya village, at the base of a rocky outcrop of gypsum anhydrites with interlayers of dolomites. The presence of caves here was suggested by V. F. Bogoslovsky (1940).

The entrances to the caves are at the level of the waterfront and are fully or partially flooded when the reservoir is filled. The width of the entrances varies from 0.5 to 1.7 m and the height—from 0.5 to 1.1 m. The main development of the caves occurs along cracks that cut deep into the massif, as well as along the coastal slope.

At the time of research (October 2020), six caves of various configurations, up to 25 m long, were noted within the site (Fig. 3.49).

Thus, in 2020, there are nine caves in the Polaznensky site, two of which—the Konets Gor and Belaya Gora—

Fig. 3.44 Plan and section of the Kulikovskaya Cave (Basharina, Gerasimova 2020)

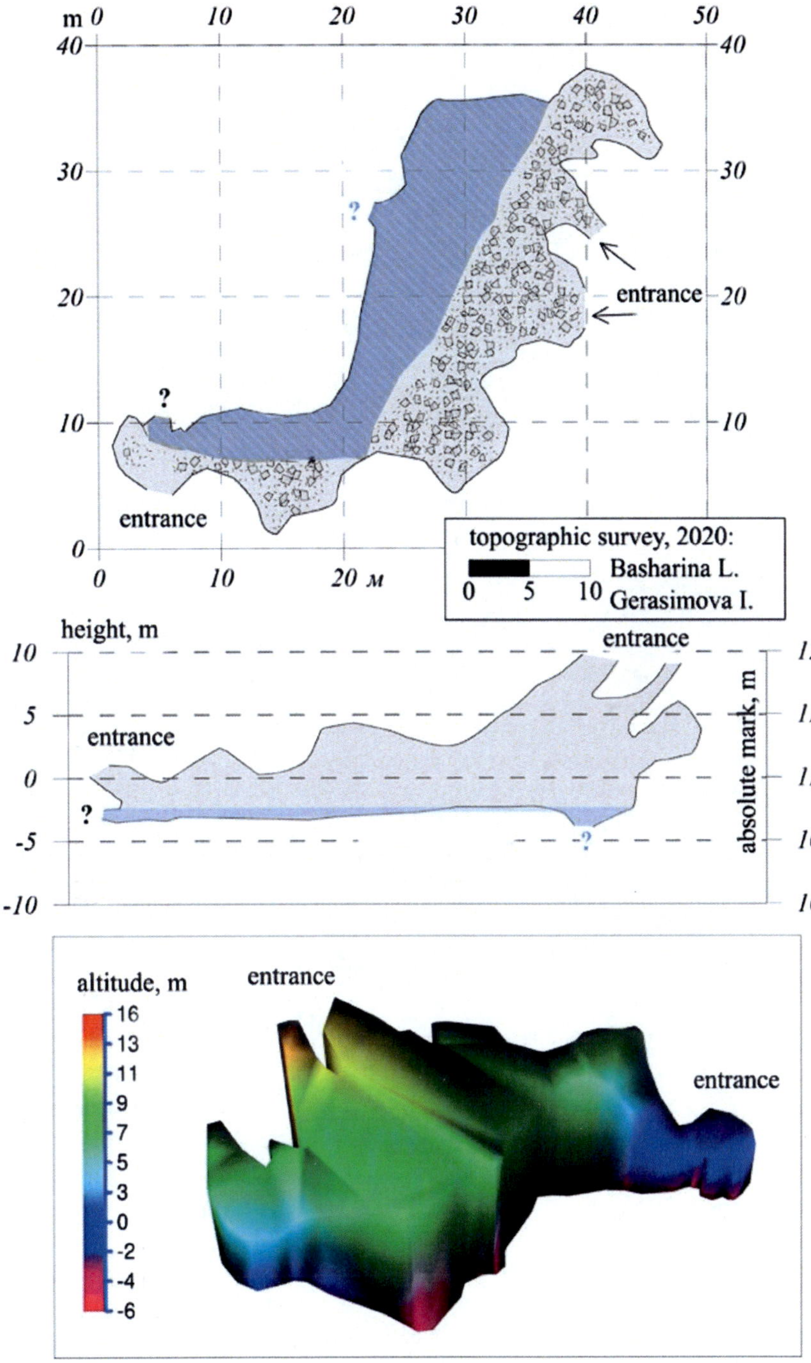

a

b

Fig. 3.45 Entrances to the Kulikovskaya Cave (*photo by* O. Kadebskaya) **a**—the main entrance; **b**—upper entrance

a

b

Fig. 3.46 Entrance (**a**) and main gallery (**b**) of the Beloy Gory Cave (*photo by* O. Kadebskaya)

located in the drainage zone of the reservoir, have significantly transformed relative to the descriptions of Bogoslovsky (1940), Pecherkin et al. (1965) and Pecherkin et al. (1981).

Existing and proposed specially protected natural areas

Existing SPNAs. Within the Polaznensky karst region, there are six SPNAs of regional and local importance with a total area of 185.5 ha (Specially Protected … 2002, 2017).

Verkhnyaya Kvazhva is a protected landscape of regional significance with an area of 1649 ha, located 8 km west of the Shemeti village, Dobryanka district. Proposed for protection in 1970, it is located on the II and III terraces of the Kama River. The largest area is occupied by sandy loam soils on deep sands underlain by karst limestones and gypsum, which causes the development of sinkholes up to 10 m deep and up to 50–100 m in diameter.

The **Durnyatskaya kettle** is a landscape natural monument of regional significance with an area of 335.7 ha, located 8 km northwest of the Peremskoye village, Dobryanka district. Proposed for protection by K. G. Butyrina in 1965. It is a classic karst flow depression with spring lakes. Permian deposits are overlain here by thin eluvial loams. There are 11 unique karst lakes in this area. Lakes Beloe and Rogalek are the deepest lakes in the Perm Region (for more details, see the description of the Nizhnekosvinsky site).

Lunezhsky Gory is a landscape nature reserve of regional significance with an area of 648 ha. Proposed for protection in 1911 by P. V. Syuzev. They occupy a narrow strip along the left root bank of the Kama reservoir between the Dobryanka town and the Polazna village. Here, sulfate rocks emerge on the day surface, rising above the water line to a height of 50 m. Within the SPNA are the most karst

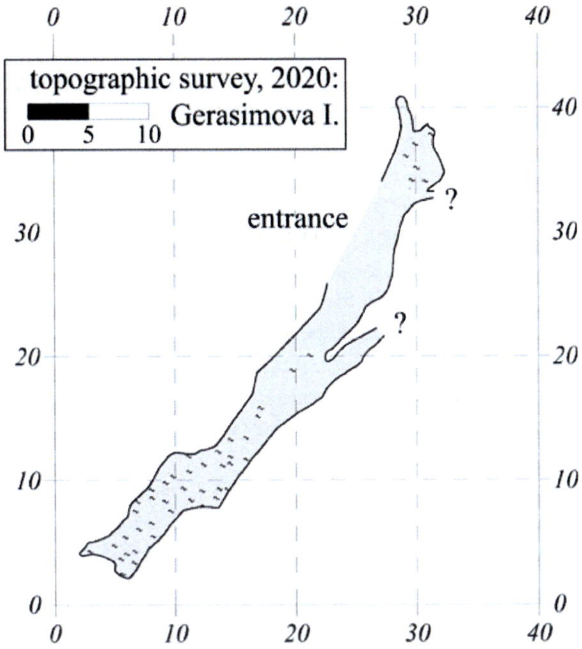

Fig. 3.47 Beloy Gory Cave plan (Gerasimova 2020)

Fig. 3.48 The entrance to the Konets Gor Cave (*photo by* O. Kadebskaya)

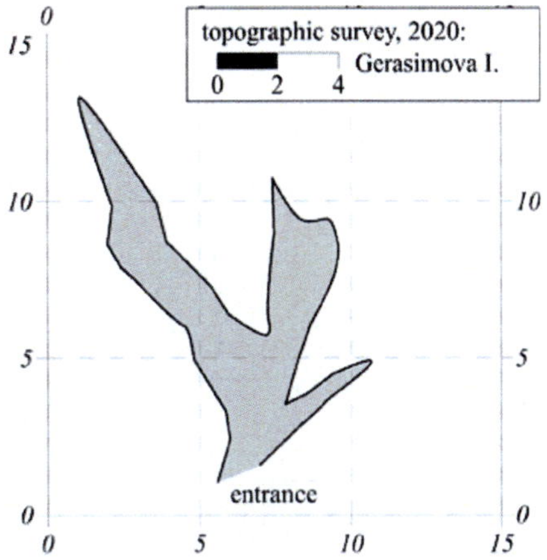

Fig. 3.49 Plan of one of the Sokoliy Kamen Caves (Gerasimova 2020)

Fig. 3.50 Specially protected natural area Sirolova Gora—gypsum outcrops on the right bank of the Chusovsky reach of the Kama reservoir (*photo by* N. Maksimovich)

areas of the Polaznensky and Lunezhsky areas of the region, where all typical forms of carbonate-sulfate karst are developed.

Sirolova Gora is a landscape natural reserve of regional significance with an area of 120 ha. It is located on the right bank of the Chusovsky bay of the Kama reservoir, west of the Kulikovo village (Fig. 3.50).

First described in 1940 by A. N. Ponomarev and proposed for protection by him together with M. M. Danilova in 1960. It is represented by gypsum rock outcrops. It is advisable to increase the territory of the specially protected natural territories by 500 m upstream to the Kulikovo village in order to include the largest Kulikovskie caves in it.

Lake Syrnik (Vask-Ivanovo) is a geological natural monument of regional significance with an area of 0.9 ha. It is located 1.5 km southwest of the Kosogor village. It is an oval-curved karst lake with flotation. The lake is 104 m long, 84 m wide, and 9.5 m deep.

Lake Shalashninskoye is a geological natural monument of regional significance with an area of 0.7 ha. It is located on the right bank of the Chusovsky reach of the Kama reservoir near the northwestern outskirts of the Shalashnaya village (Fig. 3.51).

It is a karst flowing lake 120 m long, 80 m wide, and up to 7 m deep. Formed from two merged karst sinkholes. It does not freeze in winter due to the abundant outflows of hard sulfate–calcium waters from the sides of the lake basin. A stream flows out of the lake.

Proposed SPNAs. According to the authors, the following natural monuments of regional significance were

Fig. 3.51 Karst Lake Shalashninskoe (*photo by* O. Kadebskaya)

unreasonably excluded from the list of protected areas after 2009.

Lake Bolshoye II was previously a geological natural monument, the second largest karst lake in the Perm Region (3.8 ha). Located on the left bank of the Zaozernaya River, 6 km north of the Mutnaya village. Twelve karst sinkholes were recorded at the bottom (Specially Protected … 2002).

Lake Vilvenskoye was previously a geological natural monument, located in the floodplain of the right bank of the Vilva River. It is a rounded bowl-shaped karst flowing lake 50–60 m in diameter. It has a chemically inhomogeneous composition of water at different depths.

Lake Ischezayushcheye (formerly a protected area of local significance) is a karst lake with a depth of 10.2 m.

It is necessary to restore the protected status of the previously identified objects, in addition, to submit for consideration the unique paleontological site of the Polazna section—the location of the Permian stromatolites and flora. Well-preserved remains of trunks, phylloids, and sporophylls are found in the Polazna section, located in the railway cut at the stations of Polazna and Divya, as well as in the mined-out part of the Chumkassk quarry near the Demidkovo village. Here, the first reliable finding of the sporophyll Viatcheslavia vorcutensis was made (Naugolnykh and Zavjalova 2004). In the Chumkassk quarry, there are numerous stromatolite structures of a mushroom shape up to 1 m in height (less often—3 m), which differ from the host rocks in their thin-platy structure. They belong to the spheroidal stromatolites Stratosphaerella and are characterized by a small lumpy surface. Somewhat higher in the section, among the pelitomorphic dolomites and marls, on the bedding surface, imprints of ice crystals are recorded (Geological monuments … 2009).

Economic activity and environmental pressure. The leading role in the structure of the region's industry belongs to the fuel and energy complex. Here, in 1929, P. I. Preobrazhensky discovered the first oil field in the Urals, and its production is actively carried out. The electric power industry is represented by the Kama HPP—the largest in Europe, was built in 1954 and the Kama reservoir was formed. The area is covered by a developed network of railways and highways; construction of large bridges across the Chusovskaya reach of the reservoir is underway. Sulfate karst in conditions of high technogenic load creates a number of problems in the development of this territory.

Development of oil fields. The Polaznenskoye oil field is located on the shore of the Kama reservoir and its water area. Here, since the 70s of the twentieth century, the discharge of oil-contaminated karst groundwater in the reservoir has been observed (Fig. 3.52).

The field was discovered in 1937. There are objects confined to the Tournaisian, Visean, and Bashkir-Vereian deposits in commercial operation. The volume of initial recoverable reserves is 8 million tons. As of 2009, 90 wells and 2 sidetracks were drilled, including 13 producing wells. The abandoned wells stock accounted for 68% of the total stock (Lyadova et al. 2010). The oil is of exceptionally high quality—especially light, low-viscosity, low resinous, medium-sulfur, and paraffinic. The Polaznensky karst site has a number of features caused by the development of karst, which contribute to the pollution of the first aquifer from the surface with oil products. First of all, this is the almost complete absence of surface runoff. Karst creates a significant filtration heterogeneity of the massif.

A number of researchers associate the pollution of the hydrosphere of the Polaznensky site with the ingress of oil into the aquifer during its production and transportation and others—with its upward migration (Kostarev 2004, 2015; Oborin et al. 2008).

The studies have established that the cause of the pollution of the massif, as well as the reservoir, is an oil lens with a thickness of 2–3 m on the surface of fissure-karst waters in the rocks of the Kungurian stage at a depth of 30–40 m with absolute marks of 100–108 m, which is similar in composition to the oil of the Polaznenskoye field. The most likely reasons for its formation could be crossflows from productive formations through emergency wells and the annulus of defective wells, oil discharges into sinkholes in the first years of field operation, as well as accidents on oil pipelines (Maksimovich and Meshcheryakova 2009a, b; Pikovsky et al. 2016).

The conditions of oil pollution of the massif, namely the structural and lithological-petrographic features of the massif, the physicochemical characteristics of the pollutant, the technogenic load on the study area, and others, predetermined here a special mechanism for the formation of oil pollution.

Fig. 3.52 Outputs of oil-contaminated groundwater (*photo by* N. Maksimovich)

Pollution of the hydrosphere has a rather complex mechanism, due to the significant filtration heterogeneity of the massif, the difference in the densities of oil and water, fluctuations in the level of the reservoir, the development of microbiological processes, etc.

Karsted, inclusively from the surface (up to 30% voidness), the massif was a good reservoir for oil and has a close hydraulic connection with the reservoir. As a result of bituminization of loose deposits of the coastal zone, oil was trapped in a kind of hydrodynamic trap (Fig. 3.53). Rainwater and melt water (about 280 thousand m^3 per year, taking into account evaporation per lens area) infiltrate into the karst massif and, due to the density difference, are filtered through the lens. Waters are polluted with soluble petroleum products and flow into the reservoir. The intensity of their inflow is controlled by fluctuations in the reservoir level, reaching 6–7 m per year, by the amount of precipitation, and by the intensity of snow melting; therefore, the concentration of oil products in the reservoir is extremely variable over time.

The lithology of the rocks of the massif determines their low sorption capacity in relation to the pollutant. High karst

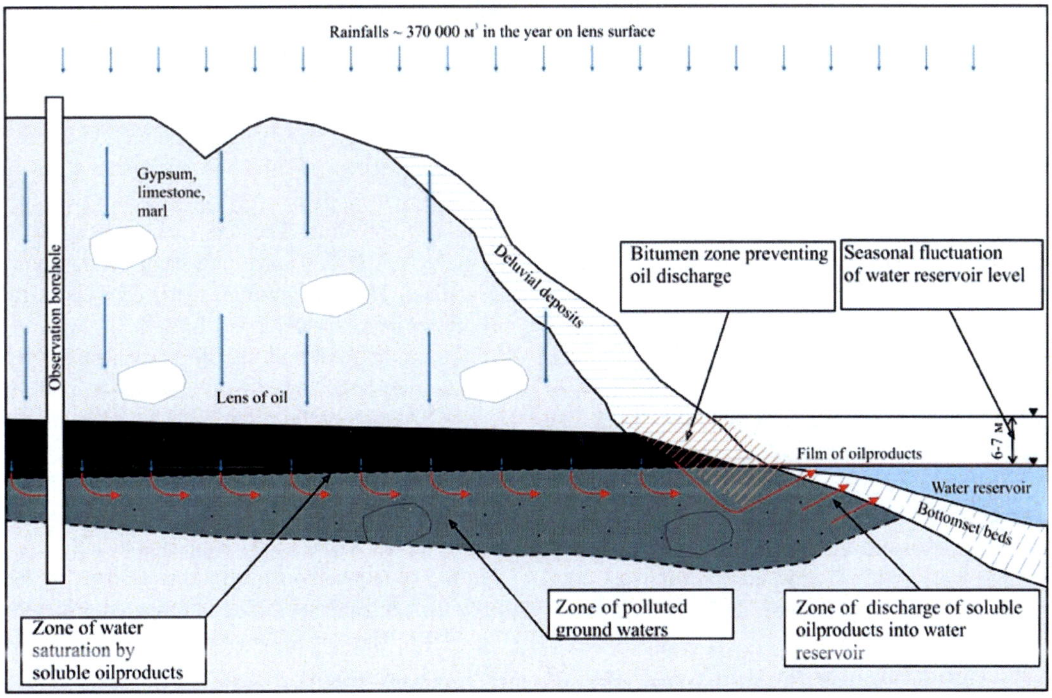

Fig. 3.53 The mechanism of pollution of the massif with oil products

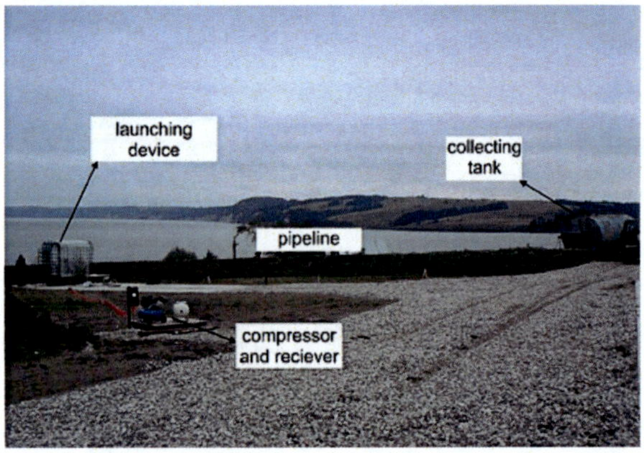

Fig. 3.54 Installation for pumping oil from the surface of fractured karst waters

Fig. 3.55 Concrete dam of the Kama HPP (*photo by* N. Maksimovich)

content and fracturing lead to rapid and unimpeded penetration of the pollutant into the depths of the massif. Karst cavities and cracks are both channels for the migration of the pollutant and its reservoirs, where oil products accumulate. Subsequently, they are partially discharged into the reservoir and act as a secondary source of pollution.

At present, groundwater pollution occurs mainly due to the carryover of dissolved oil products during filtration through the lens of rain and melt water. At the field in the coastal zone of the reservoir, there is a drainage system to intercept a part of oil products, as well as oil booms in the water area.

Based on the study of the pollution mechanism, L. N. Popov and N. G. Maksimovich developed, tested, and patented a special installation for pumping oil through existing observation wells, which allows reducing the lens thickness to 5 cm (Fig. 3.54).

The water content of the pumped oil was 3–4%. Reducing the power of the lens significantly reduces the carryover of oil products by reducing the path of water filtration through the oil. The supply of oxygen, which is necessary for the chemical and biological oxidation of oil, is enhanced. For the purification of underground waters from dissolved petroleum products, a microbiological method and regulations for the introduction of biomass into wells have been developed and patented. For this purpose, a consortium of aboriginal activated microorganisms selected from the underground waters of the field was selected. A device for monitoring water purification processes from oil products has also been patented (Maksimovich et al. 2016a, b; Maksimovich and Meshcheryakova 2009a; Maksimovich and Khmurchik 2006, 2007; Popov and Maksimovich 2008 2008; Khmurchik and Maksimovich 2007, 2015; Maximovich et al. 2005).

Hydraulic engineering. The Kama HPP, built in 1954 downstream of the confluence of the Chusovaya River into the Kama River, is currently located within the Perm city. The design head of the dam is 21 m and its length is 2.5 km. A spillway concrete hydroelectric power plant with a length of about 400 m is located in the riverbed near its steep right bank (Fig. 3.55) (Maksimovich 2006).

The zone of influence of the structure includes Permian deposits: (from top to bottom) Sheshminsky horizon, represented by interbedded mudstones, siltstones, and sandstones; Solikamsky horizon, composed of limestones, dolomites, and marls with interlayers, lenses, gypsum crystals (up to 35%). Below, there is gypsum in the anhydrite of the Irenian horizon, which is a regional aquiclude (Fig. 3.56).

Rocks in the section are characterized by heterogeneous fracturing. The geotechnical feature of the dam is the presence of readily soluble gypsum at the base, and in the sulfate-carbonate stratum, in general, an increase in the content of calcium sulfate down the section is observed.

Wells at depths of 59–73 m encountered a sulfate stratum, represented in the top by gypsum 1.6–5.6 m thick, below—by anhydrite. The rocks of the base of the hydroelectric power station have unequal gypsum content along the vertical. The least gypsum is the deposits of the Sheshminsky horizon (up to 1.3%) (Table 3.3).

Gypsum is found in the form of lenses, interlayers, individual crystals, and also in a dispersed state. Under an electron microscope, gypsum is observed in mudstone and siltstone in the form of poorly preserved elongated crystals. Less commonly, it occurs in the form of crystals up to several centimeters in size (Fig. 3.57). Its content by weight in some layers reaches 35% (Maksimovich 2006).

In the sulfate-carbonate strata, in general, an increase in the content of calcium sulfate is observed down the section (Fig. 3.58). The largest amount of gypsum is confined to dolomites, where its average content is 15%, and the smallest (3%)—to marls (Kuznetsov 1952).

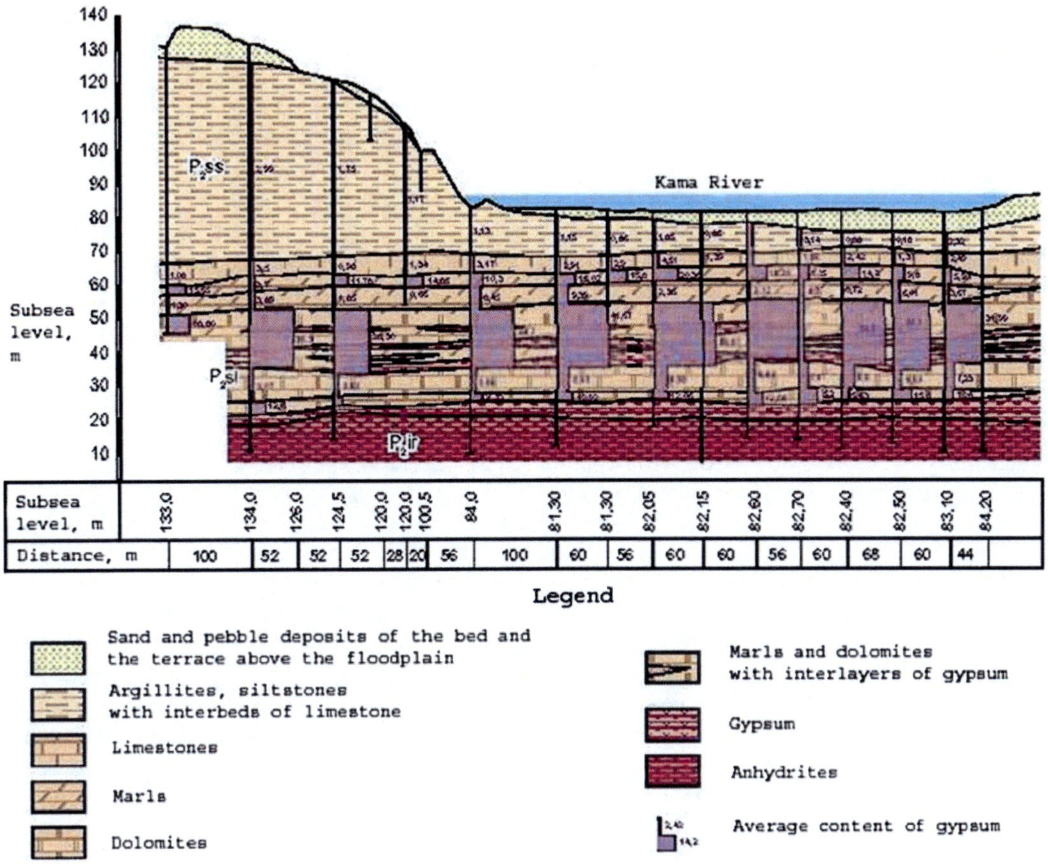

Fig. 3.56 Gypsum content of rocks along the section of the Kama dam (according to the data of surveys for construction)

Table 3.3 Gypsum content in the base rocks of the Kama HPP (Mamenko 1969)

Stratum	Thickness, m	Type of gypsum spread	Average gypsum content in % by weight
Mudstones and siltstones	5–7	Dispersed, individual crystals up to 5 m	1.3
Limestone platy	4–5	Dispersed	2.0
Dolomites	2.5–3	Small inclusions, interlayers up to 0.5 m	15.9
Calcareous marl	6–8	Dispersed, crack filler	3.1
Gypsum-bearing marls and dolomites	11–19	Interlayers, lenses with a total thickness of about 9.5 m	35.2
Clayey dolomites	9.0	Dispersed	4.9
Banded dolomites	3.5	Dispersed	12.6
Gypsum	5.0	–	100
Anhydrite	115	–	–

The most common are interlayers of fibrous gypsum with a thickness from fractions of a millimeter to 0.5 m. Their number is significant, for example, there are up to 50 layers per 10 cm of core. The thickness of the layers is maintained at a considerable distance. The contact of gypsum with the host rocks is even (Fig. 3.59a), when splitting along it, smooth surfaces are formed.

Dispersed gypsum occurs in various forms: from microscopic crystals and their intergrowths (Fig. 3.59b, c) to nests and lenses formed by randomly oriented crystals with an admixture of host rocks (Fig. 3.59d).

In the area of the dam section, carbonate-sulfate deposits are overlain by non-karsting, low-permeable rocks in the vertical direction, which prevents the development of karst.

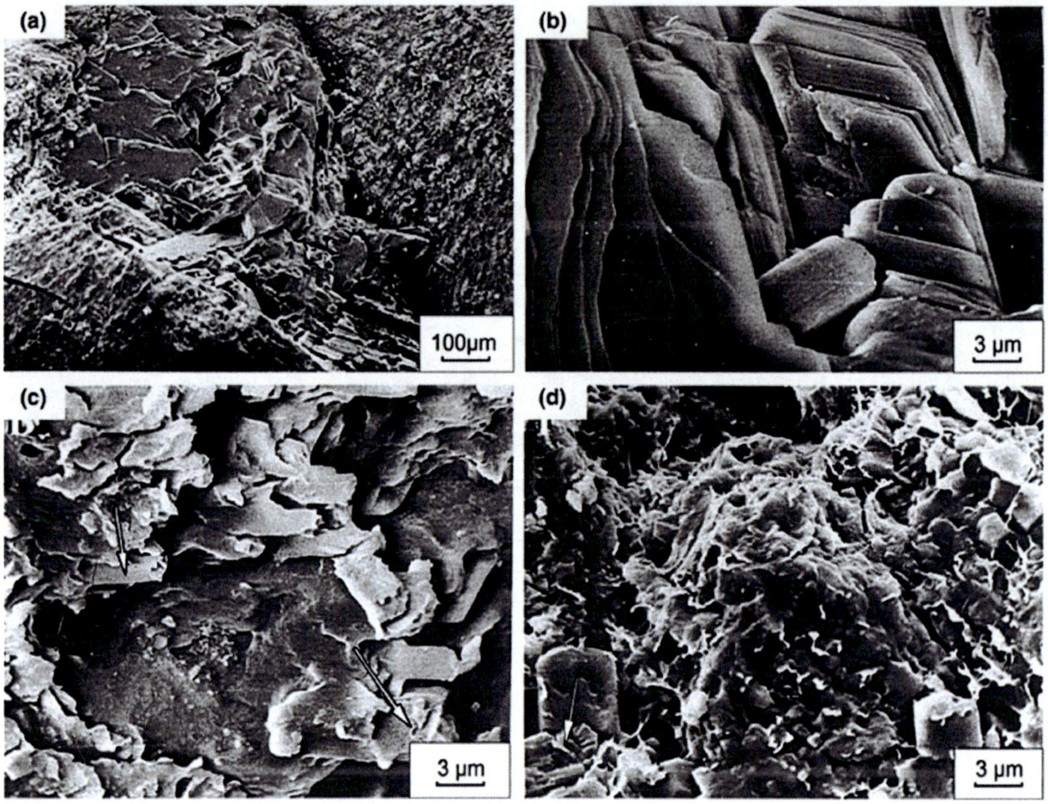

Fig. 3.57 Gypsum in the rocks of the Sheshminsky horizon. **a**, **b**—Crystal in mudstone, dispersed gypsum; **c**—in mudstone; **d**—in siltstone (shown by arrows). Cleavage perpendicular to the bedding

In the northern direction, the roof of the rocks of the Irenian horizon rises, and in the area of the Chusovskaya Strelka cape of the Kama reservoir, 400–500 m from the dam site, they lie at a shallow depth. Before the creation of the reservoir in the lower part of the steep Kama coast, under the sediments of the Solikamsky horizon, the gypsum of the Irenian horizon was exposed.

The hydrogeological situation that existed in natural conditions has changed significantly after the construction of the hydroelectric power plant and during its operation. Three stages can be distinguished: the first—before the creation of the reservoir (1954); the second—from 1954 to the end of the 70s, characterized by a change in the hydrodynamic environment and insufficiently effective operation of anti-seepage devices under new conditions; the third—from the beginning of the 1980s, when a change in hydrogeological conditions is noted after additional compaction of the cementation curtain with a silicate solution (Maksimovich 2006).

Before the creation of the reservoir, three aquifers were identified at the base of the HPP: the Sheshminsky horizon, which is characterized by hydrocarbonate, sulfate-hydrocarbonate waters with a salinity of 0.4–2 g/dm^3; the Upper Solikamsky with sulfate and chloride-sulfate waters (1.5–35 g/dm^3) and the Lower Solikamsky, the waters of which are brines with a salinity of up to 80 g/dm^3.

After the construction of the dam at its foundation, significant changes took place in the hydrodynamic and hydrochemical conditions: a change in the head of groundwater and filtration from the reservoir of fresh water at the base of the structure. A certain influence was exerted by the periodic switching on of drainage, a decrease in the effectiveness of the grout curtain over time, the introduction of a large amount of cement extraction into the massif when creating a curtain, leaks in pots, etc.

In the Sheshminsky aquifer, the area of discharge of the waters of the underlying horizons shifted towards downstream, and the desalination of groundwater began. In some areas, the dry residue decreased from 2–3 g/dm^3 in 1958 to 0.2–0.5 g/dm^3 in 1973, the sulfate ion content decreased significantly, and the hydrocarbonate ion content increased.

In the Upper Solikamsky aquifer, the levels have increased and exceeded the natural level by 3–6 m; recharge occurs both from the saline waters discharged from the Lower Solikamsky aquifer and from the desalinated waters of the underlying horizon (Oradovskaya 1962). In the Lower Solikamsky aquifer, no significant changes have occurred.

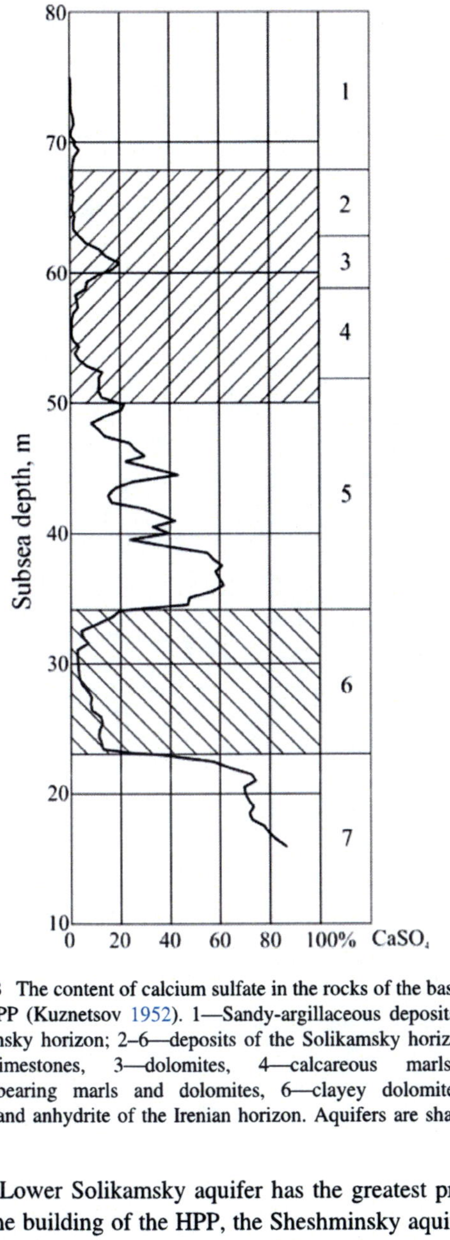

Fig. 3.58 The content of calcium sulfate in the rocks of the base of the Kama HPP (Kuznetsov 1952). 1—Sandy-argillaceous deposits of the Sheshminsky horizon; 2–6—deposits of the Solikamsky horizon: 2—platy limestones, 3—dolomites, 4—calcareous marls, 5—gypsum-bearing marls and dolomites, 6—clayey dolomites; 7—gypsum and anhydrite of the Irenian horizon. Aquifers are shaded

The Lower Solikamsky aquifer has the greatest pressure under the building of the HPP, the Sheshminsky aquifer has the smallest, that is, groundwater is discharged into the river bed.

When designing the Kama HPP, taking into account the hydrogeological features of the alignment area, an anti-seepage grout curtain and vertical deep drainage were provided. During operation, the problem of gypsum dissolution was faced.

The data of regime, hydrochemical and hydrodynamic observations showed that the processes of desalination of groundwater began to develop at the base of the hydroelectric power station, leaching and dissolution of gypsum-bearing rocks took place. In this regard, the cement curtain was additionally compacted with an oxal-aluminosilicate solution developed in the Laboratory of the Geological Faculty of Moscow State University by S. D. Voronkevich and L. A. Evdokimova, which has a high penetrating ability. The grouting effect of this solution is achieved due to the formation of a gel from a colloidal solution (sol) after its introduction into the massif by injection. The use of a silicate solution for plugging fractured soils at the base of a dam in domestic practice was carried out for the first time (Maksimovich 2006; Maksimovich and Sergeev 1983; Voronkevich et al. 1976).

After additional compaction of the curtain, its filtration coefficient decreased, and the greatest effect was achieved in the finely fractured rocks of the Upper Solikamsky aquifer, where it decreased from 0.19 to 0.002 m/day, i.e., almost 100 times.

In the Upper Solikamsky aquifer, composed of rocks with a high gypsum content (up to 35%), which is characterized by mineralized groundwater (more than 2 g/dm^3) with a significant concentration of calcium and magnesium salts, the coefficient filtration of both the curtain itself and rocks in the adjacent zone with a thickness of 5 m.

The creation of a dense curtain led to a change in the filtration pattern at the base of the dam. A decrease in the pressure after additional compaction of the curtain in the Upper Solikamsky aquifer intensified the overflow of highly saline waters of the Lower Solikamsky aquifer into it. From the point of view of the preservation of gypsum, this fact should be considered as positive, since the discharge of the waters of the Lower Solikamsky horizon into the Upper Solikamsky one, which took place in natural conditions, prevented the development of gypsum karst.

Bridge building. During the construction of a railway bridge across the Chusovaya River (Fig. 3.60), 1.5 km before its confluence with the Kama River, several wells were drilled under the abutments of the projected bridge, which exposed buried sinkholes and cavities, including under-channel voids, often with flow of karst waters. The highest karst content was observed in the area of the right bank. Here, in gypsum at a depth of 20–27 m, voids with a maximum vertical amplitude of 6–7 m were found (Fig. 3.61). To ensure the safety of the bridge, a grouting mixture was pumped into the karst cavities, consisting of loam and cement in a 1:1 ratio. It was supposed to pump 3 thousand m^3 under one support, and 6 thousand m^3 of solution under the other. The project was drawn up with a significant margin, taking into account the possibility of carrying out part of the mixture by the flow of karst waters during the injection. The actual absorption was about one third of the design (Maksimovich and Gorbunova 1958).

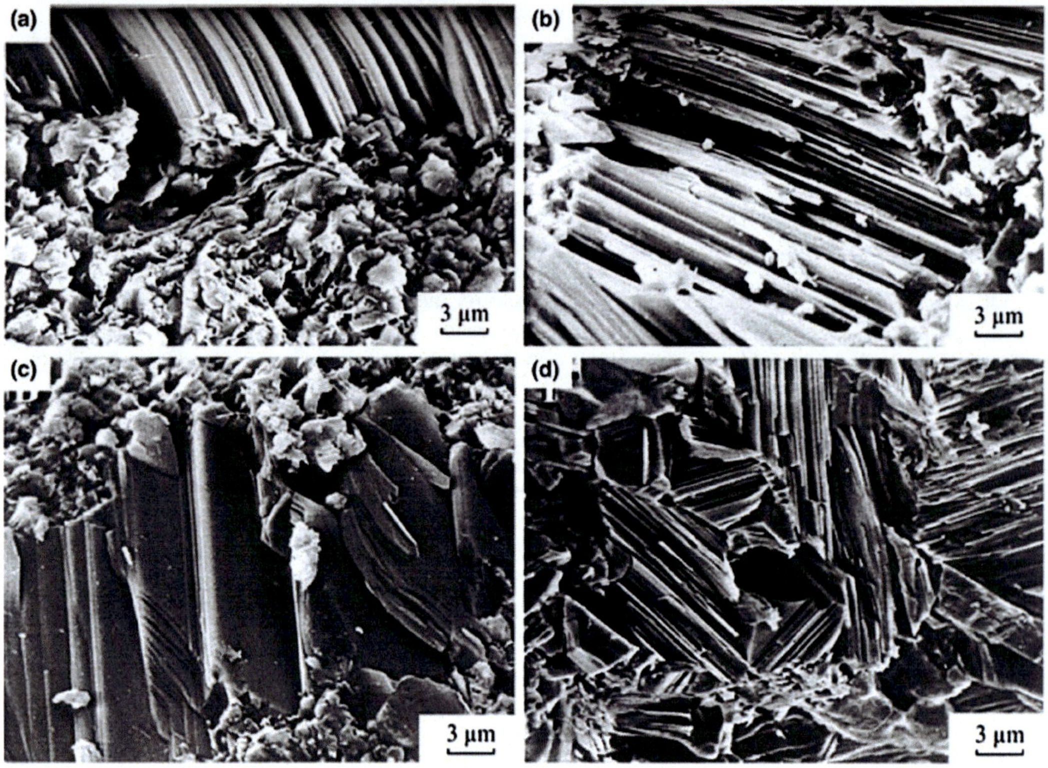

Fig. 3.59 Gypsum in the rocks of the Solikamsky horizon. **a**—contact of gypsum-selenite with dolomite marl; **b**, **c**—gypsum inclusions in dolomite limestone; **d**—gypsum nest in dolomite marl; **a**, **d**—perpendicular; **b**, **c**—spall parallel to bedding

In 1996, a two-lane road bridge 12 m wide and 1.6 km long was built across the Chusovaya River, 1.5 km upstream of the railway bridge.

During surveys in 1977 at the section of the bridge over the Chusovaya River in two wells in the channel of the Chusovaya River, two karst cavities were recorded: an unfilled one, 0.8 m high, and one filled with clay material, 1.5 m high. On the bank of the river in 1975, on the road leading to the then-existing ferry, a hole with a diameter of about 3 m was formed.

The construction of the second stage of the bridge is currently underway. During the surveys carried out in 2011 at the section of the bridge over the Chusovaya River, karst rocks of the Irenian horizon were uncovered, occurring at depths of 25.7–29.0 m, with a thickness of 22.3–23.6 m. In one of the wells in the intervals of 30.2–30.8 m and 32.7–33.1 m, unfilled karst cavities were found. Sulfate deposits are exposed at depths of 25.7–57.6 m and are represented by anhydrite, with gypsum and dolomite nests. In another well in the interval of 45.9–46.8 m, a filled karst cavity was uncovered (Utemova and Papirovaya 2015). These circumstances must be taken into account to ensure the safety of the bridge.

3.5 Lysvensky Region of Local Distribution of Sulfate Karst

The region occupies the northern part of the Sylvenskaya depression in the lower reaches of the Vilva, the Usva, and the Lysva Rivers (Chusovaya River basin) and has an area of 505 km^2 (Figs. 3.62 and 3.63). Located within the Lysva and Chusovoy districts. K. A. Gorbunova (Gorbunova et al. 1992) identified separate karst regions on this territory. However, given the similarity of the nature of karst phenomena, the authors considered it possible to define this territory as a self-sustained karst region.

The relief of the territory is hilly and ridged with average absolute elevations from 109–250 m in the west to 400–500 m in the east, which reflects the general rise of the area towards the Urals.

The main rivers of the region are Chusovaya with the main tributary the Usva River (with tributaries—the Vilva River and the Vizhai River). The valleys of the Chusovaya River and its tributaries are cut at 150–200 m, create a dismembered terrain, and abound in bedrock outcrops. Spring flood on rivers begins at the end of the second—beginning

Fig. 3.60 Railway (**a**) and automobile (**b**) bridges across the Chusovaya River, built on the karst gypsum of the Irenian horizon (*photo by* N. Maksimovich)

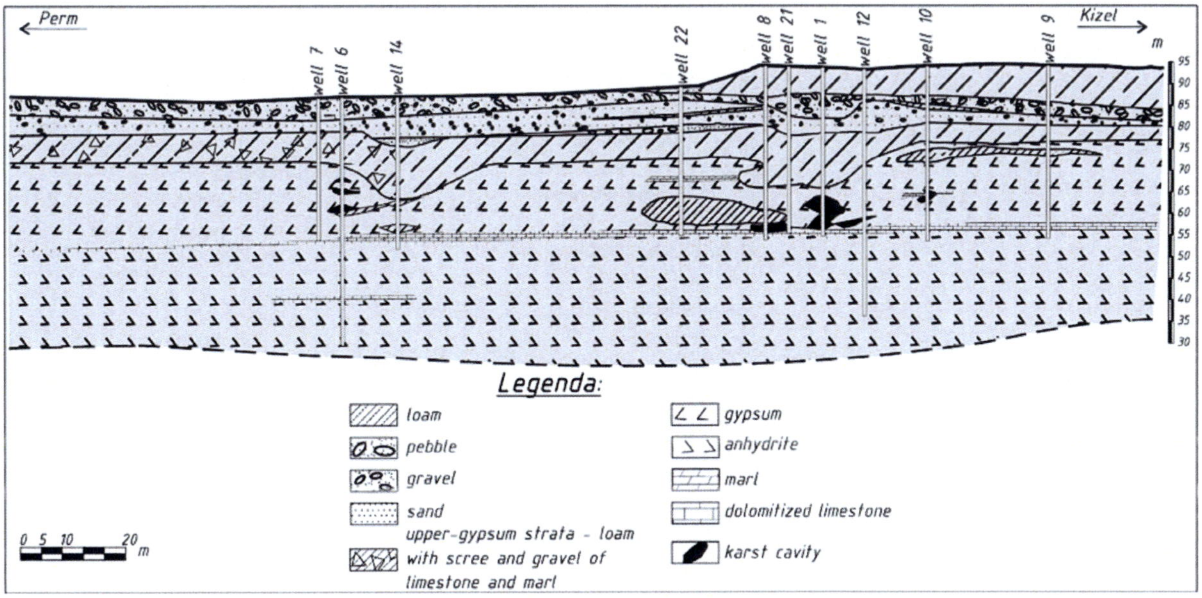

Fig. 3.61 Geological section of the right bank of the Chusovaya River at the base of the railway bridge (archived data of the laboratory of the geology of technogenic processes of the Natural Science Institute, Perm State National Research University)

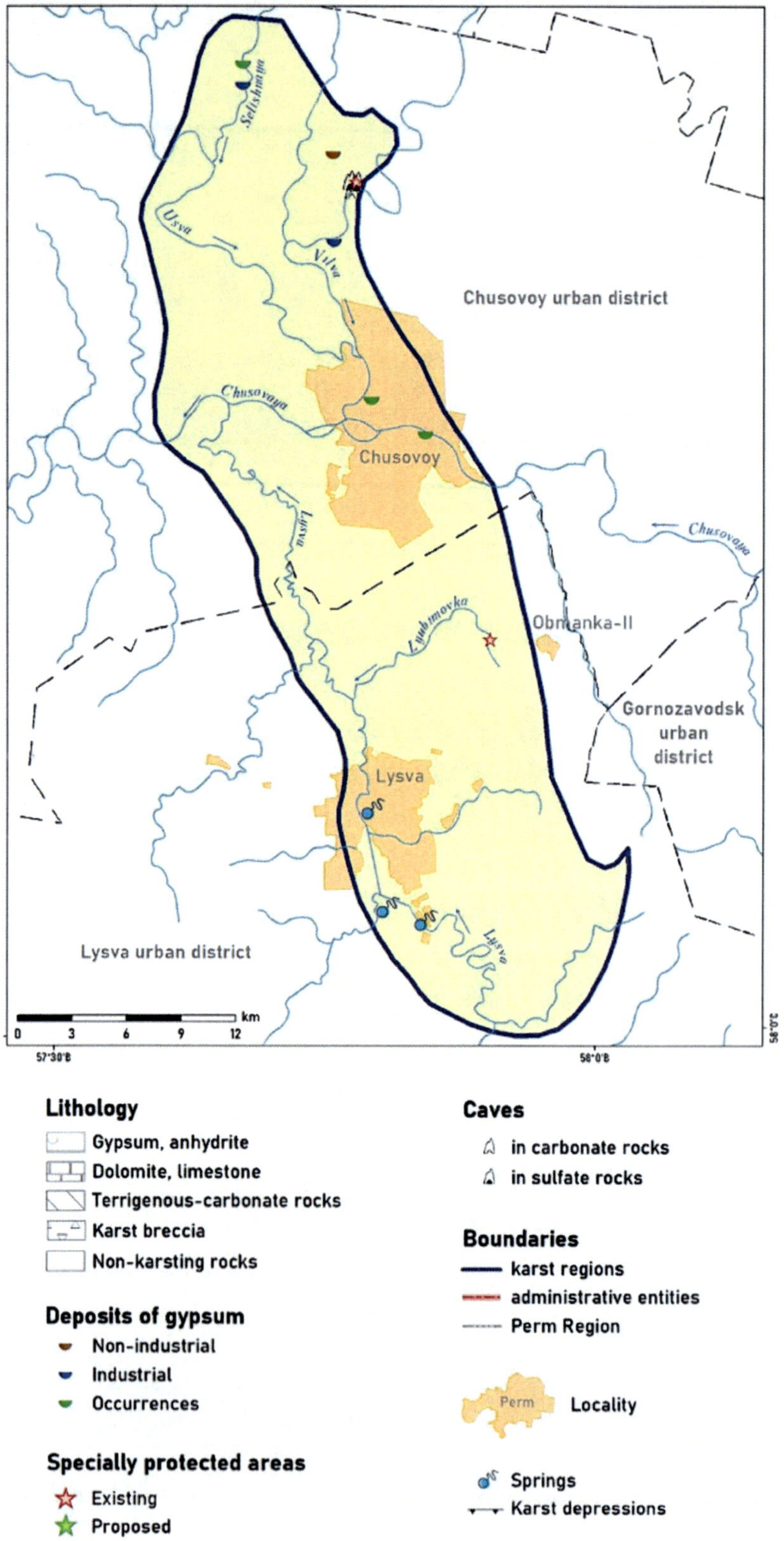

Fig. 3.62 Lysvensky karst region

Fig. 3.63 Outcrops of gypsum from Mount Opoka on the right bank of the Vilva River (*photo by* O. Kadebskaya)

of the third decade of April and ends on average in the second half of June.

Geological conditions for the development of karst. The region is located along the eastern border of the Cis-Ural trough and the West Ural fold zone. Here, the gypsum of the Leksky and Popovsky formations is the most karst. In the section, gypsum is exposed in local areas. The Lower Permian formations are represented by the Asselian, Sakmarian, Artinskian, Kungurian, and Ufimian stages. Layers and lenses of gypsum and anhydrite are found in the section of the Leksky and Koshelevsky formations. Sulfate deposits are interbedded with polymictic sandstones, siltstones, conglomerates, marls, and limestones.

From south to north, within the distribution of the Leksky formation, a decrease in the thickness of gypsum and anhydrite interlayers and an increase in the thickness of the overlying karst-avalanche deposits are observed (Snitko et al. 2017).

Hydrogeological conditions. The aquiferous Kungurian complex is represented by two horizons in the terrigenous deposits of the Kungurian stage: Koshelevsky and Leksky. The complex is developed along the eastern edge of the Cis-Ural trough. In the western direction, it plunges under the Solikamsky deposits with a roof depth of up to 150–164 m. Lithological variegation and inconsistency of water-bearing rocks, facies replacement of predominantly sandstone Koshelevsky suite in the west by clay-marl rocks of the Popovsky formation with simultaneous subsidence of the stratum leads to regular changes in hydrogeological conditions westward direction.

In the upper fractured zone, above the erosional incision, free-flow fractured groundwater is developed, feeding numerous descending springs. The depth of occurrence of fractured groundwater in river valleys is 1.84–12.0 m, and on watersheds and slopes—16–50 m.

Downstream of the drainage network, there are confined fractured formation waters. Often, pressurized waters, the depth of which ranges from 6 to 121 m, feed ascending springs. The Kungurian aquifer complex is characterized by a gradual increase in the pressure of groundwater as more and more deep aquifers are opened.

In terms of chemical composition, the underground waters of the Kungurian complex are very diverse. In the zone of active water exchange, fresh hydrocarbonate-calcium waters with a salinity of 0.1–0.54 g/dm^3 are developed, hydrocarbonate-sodium waters with a salinity of 0.11–0.68 g/dm^3 are rarely found. Sulfate-sodium, sulfate–calcium, and sulfate-hydrocarbonate-calcium (sodium) waters are quite common, the appearance of which is due to the leaching of gypsum, which occurs in the form of interlayers and lenses. Mineralization of sulfate waters varies from 0.37 to 4.1 g/dm^3. According to the degree of hardness, these are soft and moderately hard waters, and sulfate and chloride waters (with a total hardness of 11.1–100.1 meq/dm^3) are very hard (Snitko et al. 2017).

The aquiferous Asselian-Artinskian complex is confined to the terrigenous rocks of the Asselian, Sakmarian, and Artinskian stages of the Lower Permian. On the territory under consideration, it includes two aquifers: the Artinskian and the Assel-Sakmarian.

This complex comes to the surface in the forward folds of the western Urals, along the eastern flank of the Cis-Ural foredeep. The complex is composed of a rhythmically constructed terrigenous stratum, dipping to the west under younger sediments. Within the trough, it is the marginal slope of the artesian basin, being completely in the zone of hindered water exchange. The aquifers are sandstones, conglomerates, siltstones, and, to a lesser extent, limestones, separated by unstable water-resistant interlayers and mudstones. The depth of occurrence of fractured groundwater ranges from 0.7 to 84 m. The predominant type of springs is descending. The water abundance of the complex is due to the lithological composition, fracturing, the presence of faults, and bedding conditions. The flow rates of springs range from 0.05 to 60.0 l/s, wells—from 0.91 to 42.0 l/s.

In the chemical composition, fresh hydrocarbonate-calcium waters with a salinity of 0.09–0.48 g/dm^3 are predominantly developed. Mixed hydrocarbonate-sulfate–calcium and sulfate–calcium, sulfate-sodium waters are extremely rare. The salinity of these waters varies from 0.29 to 3.04 g/dm^3. In the zone of difficult water exchange, salty sulfate and chloride waters are developed.

Karst springs in the area of the Opoka Mountain are fed by the Opoka River, the runoff of which is absorbed in some areas, as well as infiltration of atmospheric precipitation, and have a salinity of 0.9–2.2 g/dm^3. The total flow rate of the springs ranges from 5 during low-water periods to 56 l/s during floods.

There are three known deposits of sulfate waters in the region.

The site "Lesnaya Voda" is located on the territory of the sanatorium-preventorium "Zorka", 7.5 km northwest of the city of Lysva. Mineral waters of sulfate magnesium-calcium composition with a mineralization of 2.35–2.66 g/dm^3 were discovered in the interval of 100–200 m. Mineral waters are assigned to the XII group of sulfate magnesium-calcium waters of the Kazan type.

Rest House "Sokol": a well with a depth of 70 m was drilled in 1990. It is located east of the Lysva town on the right bank of the Nevidimka River, in the vicinity of its mouth. An aquifer with mineral water is localized in the interval of 50–70 m in cavernous silicified limestones of the Koshelevsky suite. The flow rate is 4 l/s, and the water is sulfate magnesium-calcium.

Kamennolozhskoe (Lysvenskoye) groundwater deposit is located 2 km south of the southern outskirts of the city of Lysva, on the right bank of the Lysva River. Groundwater has a mineralization of less than 1 g/dm^3. The water is mainly hydrocarbonate-calcium and sulfate-hydrocarbonate-calcium. The abundant freshwater aquifer is enclosed in the fissured Kungurian rocks. Sources of constant replenishment of reserves are the inflow of fractured waters with a flow rate of about 36.9 thousand m^3/day and the Lysva River, the infiltration of which is up to 21.05 thousand m^3/day (Snitko et al. 2017).

Karst phenomena. On the Opokinsky site in the gypsum of the Kungurian stage, sinkholes, kettles, karst ravines, the disappearing Opoka River, karst springs, and caves are observed. In the Chusovsky site, the Kungurian gypsum is karsted, occurring among the carbonate-terrigenous rocks in the thrust zone. Sinkholes, dolines, and underground cavities are characteristic (Gorbunova 1977).

Since 1950, ten sinkholes with typical sizes up to 3 m in diameter and up to 2 m in depth have been recorded on the territory of the region. The largest one occurred in 1965 in the city of Chusovoy and was 7 by 8 m in size with a depth of 9 m.

According to the archival data of the Mining Institute of the Ural Branch of the Russian Academy of Sciences and information obtained from speleologists, four caves in gypsum and anhydrite of the Opoka Mountain are known in the region: Nisha, Untitled Cave, Malaya, and Bolshaya Opokinskie. The largest was Bolshaya Opokinskaya, 188 m long and 20 m deep (Caves of the Volga 2010; Agashkov and Trushnikova 1965). Due to the high rate of dissolution of sulfate deposits, caves evolve rapidly.

The **Nisha Grotto** is located on a rocky outcrop above the Vilva River. The absolute elevation of the entrance is 136 m. It has a width of 11 m and a height of 5 m. The length of the grotto was previously 7.5 m, after the destruction of the entrance part, the length of the cavity deep into the massif is no more than 3 m (Fig. 3.64).

a *b*

Fig. 3.64 Landfalls at the entrances to the Bolshaya Opokinskaya Cave (**a**) and the Nisha grotto (**b**), 2020 (*photo by* O. Kadebskaya)

Fig. 3.65 Rock outcrop and entrance to the Malaya Opokinskaya Cave (*photo by* O. Kadebskaya)

The **Bolshaya Opokinskaya Cave** is located on the slope of the largest karst basin. Before the collapse of the entrance part, one could get into the cave through a narrow ponor, in which the Opoka River was swallowed. Currently, the water goes into the dam, and the entrance to the cave is accessible only from the slope of the karst basin. The entrance to a large entrance grotto, measuring 15 by 20 m, is closed by a blockage. The length of the cave is 22 m and the amplitude is 7 m.

The **Malaya Opokinskaya Cave** is located in a chain of large depressions in the rear seam of the ancient terrace. The entrance is located on the side of the sinkhole and has a triangular shape (Fig. 3.65). The cave is a horizontal tunnel 1.2 m high and 1.5–2 m wide. At the end, the 11 m long tunnel after the blockage expands to a small grotto 5.5 m long and 2.5 m wide with a height of up to 3 m. In the vault of the grotto, there are filled organ pipes. During the survey, the cave in the entrance part was filled with ice. The total length of the cavity is 33 m and the amplitude is 8 m.

According to the results of the 2019 survey carried out by the “Speleodays” group, Malaya Opokinskaya is currently the largest cave in the area (Fig. 3.66).

Existing specially protected natural areas. Within the Lysvensky karst region, two natural monuments of regional significance with a total area of 44.5 ha have been identified.

Angydrit is a geological monument of regional significance with an area of 27.5 ha, located 4.3 km northwest of the Obmanka-II village, Lysva district (Fig. 3.67).

The slopes of the unnamed stream, a tributary of the Lyubimovka River are intensively karst. There are large numbers of sinkholes, ditches, and niches. Gypsum in the outcrop has a pinkish tint, there are interlayers of selenite and segregated gypsum with crystals up to 2 cm. Birch-spruce forests grow on the rocky outcrops of sulfate rocks (Special protected … 2017).

Opoka is a landscape monument of regional significance with an area of 17 ha. The isolation of the Opoka mountain

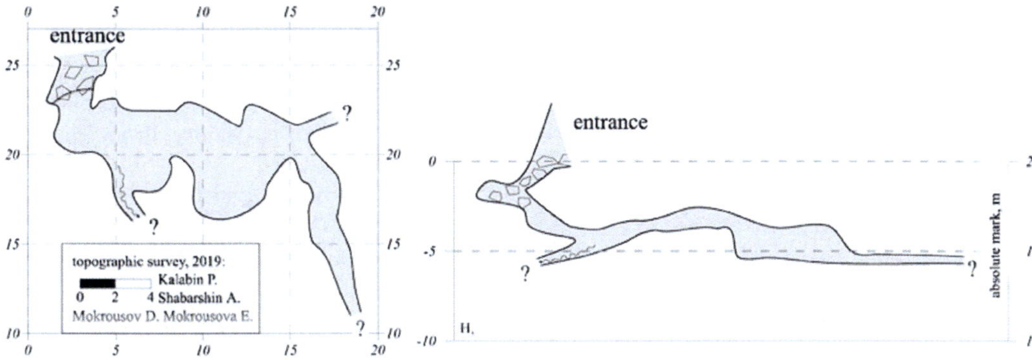

Fig. 3.66 Plan and section of Malaya Opokinskaya Cave (Kalabin, Shabarshin, Mokrousov, Mokrousova 2019)

Fig. 3.67 Outcrops of gypsum in the SPNA Angydrit (*photo by* O. Kadebskaya). **a**—Gypsum outcrop in the side of the stream; **b**—outputs of pink plaster; **c**—selenite veins between destroyed interlayers of dolomite in gypsum

massif leads to the fact that it is a standard of manifestations of sulfate karst in miniature. It has its own area of absorption of surface waters of the Opoka River and an area of discharge of karst waters—springs at the eastern foot of the Opoka Mountain. The main karst forms are sinkholes. More than 60 depressions up to 50 m in diameter have been recorded. Three karst basins are more than 100 m in diameter and were formed due to the confluence of several sinkholes. In the largest northeastern basin, 54 m deep, the waters of the Opoka River are absorbed in a ponor. Earlier, the waters of the river went under a rocky outcrop of gypsum 40 m high. Organ pipes were detected on the sides of sinkholes.

Economic activity and environmental pressure. The leading industries in the structure of the economy of this region are ferrous metallurgy, mechanical engineering, metalworking, building materials industry, light, and food industries. There are enterprises in fuel and energy complex, logging, and timber processing.

Most of the territory is occupied by the Lysvensko-Chusovskoy industrial and economic region, which is characterized as territories with an unfavorable state of environment. Extensive geochemical anomalies with halos and streams with a high content of various contaminants have been established. The most dangerous chemical elements forming technogenic anomalies are cadmium (up to 50.0 MPC), beryllium (up to 5.0 MPC), lead (up to 3.3 MPC), phosphorus (up to 3.5 MPC), zinc (up to 1.8 MPC), chromium (up to 5.3 MPC), copper (up to 1.4 MPC), nickel (up to 1.3 MPC), manganese (up to 2.7 MPC), titanium (up to 2.6 MPC), and barium (up to 1.7 MPC) (Snitko et al. 2017), which must be taken into account in the ecological assessment of the territory, since an increased migration of pollutants is noted in the areas of karst development.

The Vilva River is located in the zone of influence of the outflows of the liquidated mines of the Kizel coal basin. The river has consistently high concentrations of iron (up to 426 MPC), aluminum (up to 33 MPC), manganese (57 MPC), periodically—beryllium, cadmium, and nickel (Maksimovich and Pyankov 2018). Unfortunately, the influence of the outflow of such waters on the Opoka landscape monument has not been studied.

There are three known **gypsum deposits** in the region—Selishchenskoye, Opokinskoye, and Chusovskoye. All deposits are confined to the Leksky formation of the Kungurian stage of the Lower Permian. The strip of outcrops of the formation was mapped in the basin of the lower reaches of the Vilva and the Lysva Rivers (Snitko et al. 2017).

Fig. 3.68 Outputs of gypsum within the Opokinskoye deposit (*photo by* O. Kadebskaya)

The largest of them, the Selishchenskoye field, explored in 1941, is located in the Chusovoy region, 4.5 km northwest of the Utes railway station, on the right bank of the Selishnaya River. It is confined to the lagoon deposits of the Leksky formation of the Kungurian stage of the Lower Permian. The productive stratum of gypsum occurs among calcareous-clayey sandstones and siltstones. The deposit is lenticular, its thickness reaches 17.8 m (on average 6.1 m). Overburden rocks are represented by loams and clays with an average thickness of 1.7 m. The karst content of the gypsum strata reaches 28%. Gypsum is presented in three varieties: dense, reticulate, and thin-layered.

The Opokinskoye field is located in the Chusovoy district, at the mouth of the Opoka River, which flows into the Vilva River. An anticlinal fold is noted within the deposit, with the wings dipping to the southwest and northeast. Gypsums emerge in outcrops along the right bank of the Vilva River for 1.5 km (Fig. 3.68).

The gypsum-anhydrite strata submerge in the southwestern direction (along the Vilva River) with an average thickness of 25 m. In the central part of the deposit, the thickness of gypsum reaches 80 m. The form of the deposit is lenticular. The most promising part of the field has a length of 700 m and a width of 300 m and an area of 12 ha. The useful stratum is represented by gypsum of white, grayish-white, gray, dark gray, bluish-white, and pinkish-white color, lamellar, granular with a varied network of marly-clay inclusions. Monolithic anhydrites up to 70 m thick lie below the gypsum.

3.6 Nizhnesylvinsky Region of Sulfate and Carbonate-Sulfate Karst

Nizhnesylvinsky region covers an area of 2,191 km^2. Administratively, it is confined to the Berezovka, Chusovoy, Dobryanka, Kishert, Kungur, and Perm districts of the Perm Region, located in the lower reaches of the Sylva River (Fig. 3.69).

Geological conditions for the development of karst. The region occupies the eastern edge of the East European platform on the border with the Cis-Ural trough and is confined to the Bym-Kungur depression. Large structures are complicated by the Mazuninskaya and Veslyanskaya swell-like zones, within which local structures of smaller sizes are distinguished (Gorbunova et al. 1992).

Deposits of the Kungurian stage and rocks of the Neogene and Quaternary systems take part in the structure of the sedimentary cover of the region (Fig. 3.70). The Kungurian stage is divided into two horizons: the Philippovsky and the Irensky ones.

The Philippovsky horizon is widespread in the east of the region and is composed of layered dolomites and dolomitized limestones with a thickness of 50–80 m. The rocks contain a significant admixture of clay material. There are lenses of gypsum of low power.

The Irenian horizon is represented by three sulfate and two carbonate suites.

The Ledyano-Peshcherskaya suite is exposed in the lower parts of the slopes of the Sylva, the Babka, and the Shakva River valleys. The layer is dominated by granular-crystalline, cryptocrystalline, nodular, layered-wavy anhydrites, gypsum is less common.

Gypsum predominates in the near-slope parts occur in places of intense fracturing, as well as near contacts with carbonate rocks. The thickness of the layer is 35–40 m. The markers of the layer are borates, often in the form of nodules, represented by calcium borosilicates (howlite and studenicite).

The Nevolinskaya suite is mostly lined with dolomites and dolomitized limestones, gray, yellow-gray, and dark gray, sometimes with an interlayer of oolitic dolomite up to 1 cm thick. In the middle part of the layer, there are layers of gypsum up to 6.5 m thick and interlayers of clay up to 0.1 m

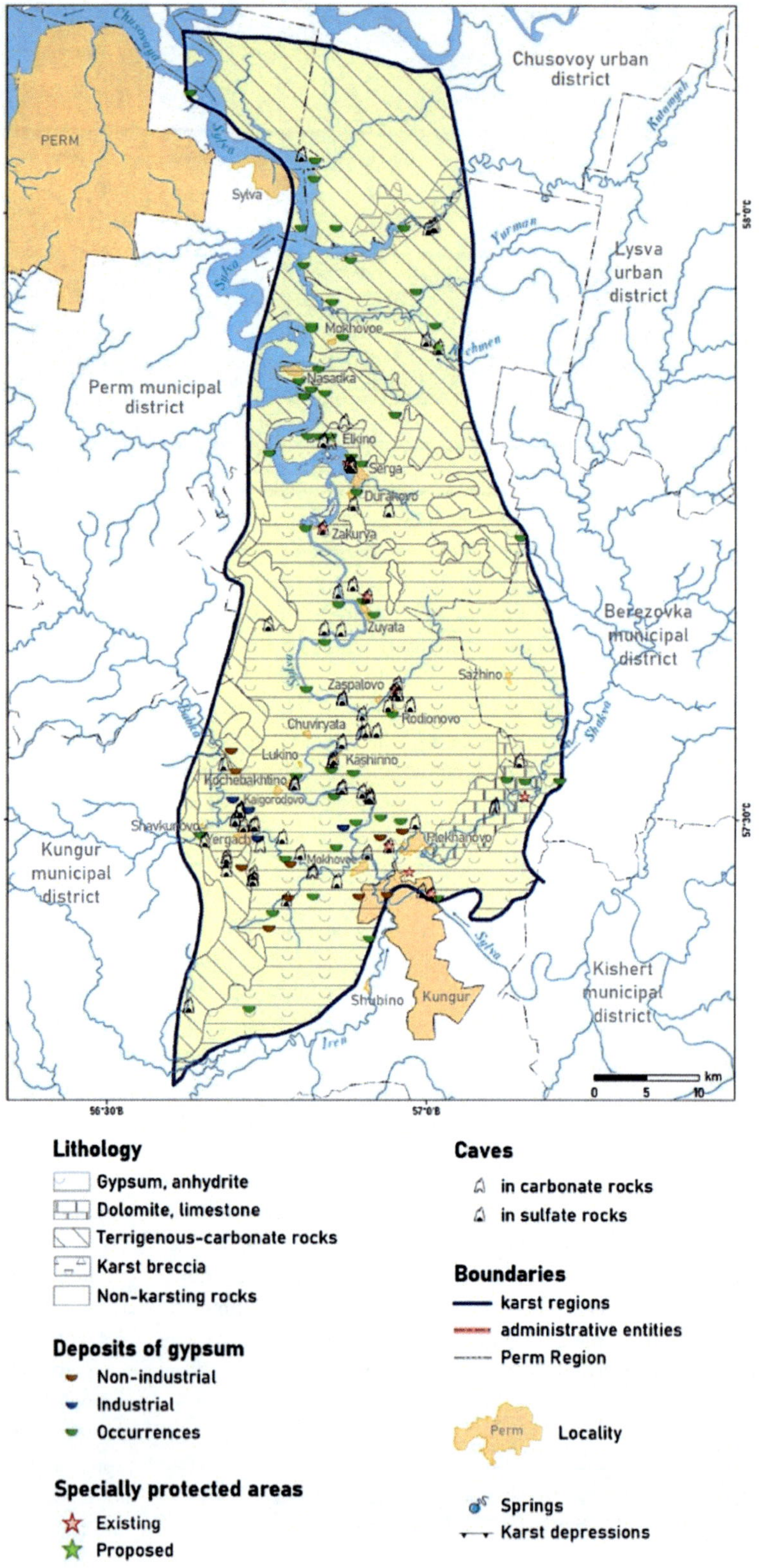

Fig. 3.69 Nizhnesylvinsky karst region

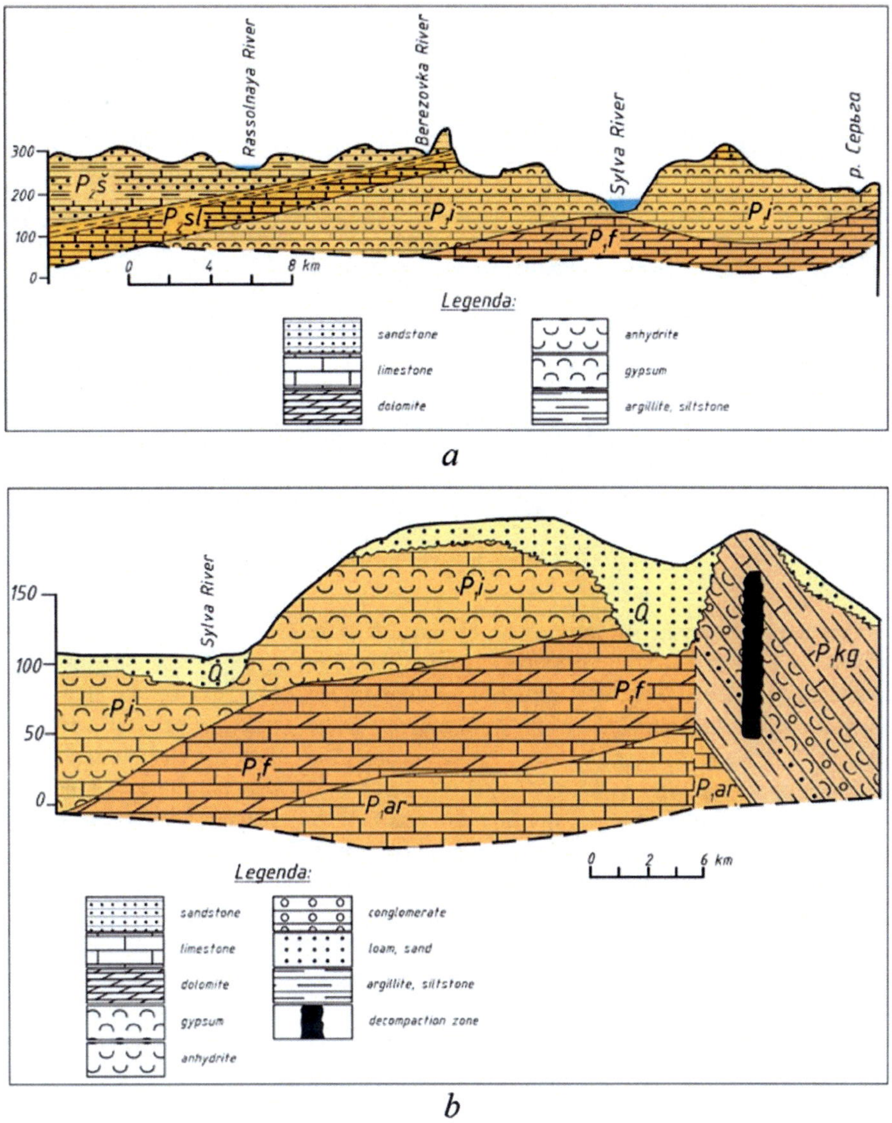

Fig. 3.70 Geological conditions for the development of karst in the Nizhnesylvinsky region: **a**—the western part (Gorbunova et al. 1992); **b**—eastern part (Lavrova 2018)

thick. Limestones of dolomitized and secondary dolomites contain a diverse fauna complex. The thickness of the layer ranges from 7.5 to 15.4 m.

The Shalashninskaya suite is represented by granular and cryptocrystalline gypsum with an irregularly wavy texture and gray, clayey, pelitomorphic, massively bedded dolomites. The layer is heavily destroyed due to karst processes and has been preserved as separate islands. The layer is a marker of the studenicite mineral found in the core of a well drilled on Ledyanaya Mountain in 2014.

The Yelkinskaya suite is represented by dolomites, dolomitized limestones, gray, oolitic, pelitomorphic, detrital, and organogenic-detrital limestones with various fauna. The outcropping of the rocks of the layer is described near the Yelkino village. The thickness of the layer ranges from 1.5 to 10.5 m (Fig. 3.71).

The Demidkovskaya suite is represented by grayish-white gypsum. Numerous outcrops of the layer 25–30 m thick occur along the banks of the Sylva River. Its sediments contain a boric mineral—probertite, the manifestations of which were described for the first time in the section of the Deikovo gypsum deposit (Kungur district) in 2013.

The Tuysky suite is represented in the lower part by white and light gray crystalline limestones, locally silicified, and fine-grained dolomites with a total thickness of 12–13 m. The layer contains an asbestos-like mineral—palygorskite, which was discovered in the Kama region in 1960 and is a reliable marker of these deposits.

Fig. 3.71 Outcrops of rocks of the Yelkinskaya suite on the right bank of the Sylva River near the Yelkino village (*photo by* N. Maksimovich)

The Lunezhsky suite is represented by grayish-white gypsum and bluish anhydrite, sometimes with an admixture of dolomite material. The most complete sections of the layer are known on both banks of the Babka River, in the vicinity of the Yergach village and the Kylasovo village. The thickness of the layer ranges from 34 to 65 m.

In the Shakva River basin, the Irenian horizon is faciesly replaced by the Popovsky formation of thin-plated marls, sandstones, clayey limestones, mudstones, and siltstones with lenses and interlayers of gypsum and rock salt. To the north of the Serga village, on the watersheds, it is overlain by rocks of the Solikamsky horizon of the Ufimian stage, in the lower part by dolomites, sometimes silicified, with interlayers of clay and selenite; in the upper one—fine-grained limestones, interbedded with platy marls, clays, siltstones, and sandstones. Olkhovskaya karst breccia is less common than in the Kishertsky karst region. The Neogene sediments of the lower part of the strata are karst-avalanche, composed of grit, rubble, and blocks of carbonate and sulfate rocks with an admixture of sandy-clay material and karst breccia up to 40–50 m thick.

Quaternary deposits of the upper part of the strata are represented by alluvial, diluvial, alluvial-deluvial, and proluvial deposits up to 40 m thick.

Hydrogeological conditions. The main waterway of the region is the Sylva River. Within the Kungur town, the Iren and the Shakva Rivers flow into it, which significantly affects its hydrological regime. This river knot was formed at the beginning of the Neogene period mainly under the influence of tectonic movements and karst and, to one degree or another, continued to form in the Quaternary period. The characteristics of the Sylva River are presented in the description of the Kishertsky karst region (see Sect. 3.7).

River waters, especially during spring floods, have an impact on the development of karst processes. Non-freezing openings on the rivers indicate the presence of powerful outcrops of karst waters. The karst waters of sulfate rocks feed the river waters, which leads to an increase in their mineralization. The increased water availability of the region is typical for zones confined to tectonic disturbances characteristic of the right bank of the Sylva River on the western outskirts of the Kungur town, the watershed of the Sylva and the Shakva Rivers, and the middle and lower reaches of the Babka River.

The orientation of regional fracture-ruptured structures suggests the possibility of localized overflows of groundwater in accordance with the direction of the regional runoff, caused, in this case, by the direction of the fall of rocks of the sulfate-carbonate complex from the northeast direction to the west, southwest, and northwest. The assumption about the hydrogeological activity of fracture-rupture structures is confirmed by their spatial coincidence with the areas of localized discharge of groundwater into the alluvium of the Sylva River valley.

The Nizhnesylvinsky karst region is characterized by a complex distribution of groundwater. In the area of distribution of the gypsum-anhydrite complex, isolated karst streams are developed (the area of sporadic distribution of the waters of the Irenian deposits), and in the limestones and dolomites of the Philippovsky horizon and the Artinskian stage, there is a zone of continuous saturation of voids with water—the Artinskian—Philippovsky aquifer complex (Hydrogeology of the USSR 1972).

The absence of regional aquicludes, different lithological compositions, and fracturing of water-bearing rocks contribute to their hydraulic connection and cause different hydrodynamics and chemical composition of groundwater. There is a close hydraulic connection between surface and ground waters.

Groundwater is confined to Quaternary alluvial sandy-pebble and loamy deposits, Irenian carbonate-sulfate, and Artinskian-Philippovsky carbonate strata of the Permian age. In the crest of the swell, the carbonate deposits of the Philippovsky horizon are aquifers. As they sink to the west, they acquire the role of a regional aquiclude, on which underground waters of karst breccias and fractured karst gypsums and anhydrites of the lower Irenian horizon are formed.

The Tuysky limestones, not disturbed by karst, represent a confining layer on which the fractured-karst waters of the Lunezhsaya suite of gypsum are formed. Where they are raised above the river bed, the Lunezhsky gypsums are usually waterless, and in the underlying layers, isolated karst streams are noted. The thin Nevolinskaya and Yelkinskaya dolomite layers are almost universally disturbed, and, only in some places, they play the role of aquicludes.

The chemical composition of the Sylva River is subject to seasonal changes, while the waters have a constant calcium-hydrocarbonate-sulfate composition. The lowest values of macrocomponents, salinity 243–384 mg/dm^3, and pH 7.1 are typical for the spring period, which is associated with an increase in the role of melt water in the river feeding at this time. In the autumn period, the waters of the Sylva River acquire a slightly alkaline character, pH 8.1, mineralization rises to 547 mg/dm^3, which is associated with the replenishment of the Irenian horizon by groundwater. This is due to the close hydraulic connection between the waters of the Philippovsky and Irensky horizons and the Sylva River. Dolomites of the Philippovsky horizon play the role of barrages, preventing the movement of waters across the strike of the rocks, and contribute to the flow of the Sylva River. In winter, groundwater outflows from the Philippovsky limestones and dolomites can be observed on the banks (Kazantseva et al. 2020).

The waters of the Shakva River are slightly alkaline (pH 7.6–8.2); they are also characterized by seasonal variability of the chemical composition. In 2017, the river waters had a constant calcium-magnesium-hydrocarbonate-sulfate composition, which in the spring of 2018 changed to a calcium-magnesium-sulfate-hydrocarbonate composition.

The isotopic composition of river waters is related to the isotopic composition of atmospheric precipitation. Correlation analysis showed that there is a moderate relationship between these indicators. This is due to the fact that, during the year river waters have, in addition to atmospheric, and underground water. Observations of the annual cycle of the isotopic composition change made it possible to calculate the time elapsing from the moment of sediment absorption into groundwater to its discharge from karst springs into the Shakva River. With a shift in the values of $\delta^{18}O$ and δ^2H for river waters by 2 months relative to atmospheric precipitation, the correlation for the Shakva River becomes strong (Kazantseva et al. 2020).

Changes in groundwater salinity of the Irenian horizon are seasonal but do not lead to a change in the chemical type of water. The water has a constant calcium-sulfate composition. Mineralization varies from 1846 to 2260 mg/dm^3.

The waters of the Philippovsky horizon are slightly mineralized (400–600 mg/dm^3), but at the contact with the Irenian gypsum, mineralization increases to 1500–3600 mg/dm^3. In the winter-spring period, the waters of the Philippovsky horizon have a calcium-sulfate composition, in the summer-autumn period, they are calcium-magnesium-sulfate-hydrocarbonate and calcium-sodium-magnesium-sulfate-hydrocarbonate (Kungur Ice Cave … 2005; Kataev and Kadebskaya 2010). When the water level on the Sylva River rises, the waters of the Lower Permian carbonate series are backed up, and, through the fractured massif, they enter the overlying Irenian horizon, which leads to a change in their salinity.

In the zone of vertical descending circulation, the water is almost completely saturated with salts, the mineralization of water in cave streams and lakes is, as a rule, 2.0–2.2 g/dm^3. The formation of cavities in the sides of the valleys occurs mainly due to low-mineralized river waters invading the karst massif during the spring flood, or/and underground, low salinity, from neighboring sediments. The depth of flood waters penetration into sulfate massifs reaches 1.5–2 km.

Karst phenomena. Karst occurs mainly in gypsum and partially in carbonate rocks of the Irenian horizon. It belongs to the bare and choked types, in large areas to the subeluvial, suballuvial—in the river valleys, covered—under the Solikamsk rocks (Gorbunova et al. 1992).

Surface karst forms are represented by depressions of various shapes and sizes and outliers. The density of depressions increases in ravine valleys and decreases with approaching watersheds. Numerous sinkholes are found in the strip of gypsum-anhydrite rocks. In some places, the conditional density of craters reaches 85 forms/km^2 (Ledyanaya Mountain). The number of craters decreases with distance from river valleys and dens. Nevertheless, it is not easy to find an area of 2–4 ha in watersheds where there are no sinkholes at all. The dimensions of the craters range from 2–3 m to 100 m or more in diameter at a depth of a fraction of a meter to 15–20 m (Kataev and Kadebskaya 2010).

By morphological features, saucer-shaped and cone-shaped funnels are more common. Well-shaped depressions in the place of recent sinkholes are present in a single quantity. Open holes and outcrops of bedrock in craters are quite rare. Swampy sinkholes are developed, as a rule, in watersheds.

Fig. 3.72 Lake in a sinkhole in the former Sapovo village (*photo by* N. Maksimovich)

Along with permanent water bodies, temporary lakes are widespread, which are formed during the period of snow melting and remain for several weeks. At the beginning of May, the surfaces of ancient terraces are literally dotted with such lakes-saucers, in some of them water remains until the end of summer. Karst lakes are of great importance in the life of the local population, the water supply in them sometimes determines the location and even the size of the settlements (Fig. 3.72).

Below the Serga village, the Sylva River is located in the backwater zone of the Kama reservoir, which led to the activation of karst processes in the coastal zone composed of sulfate rocks. The main forms on coastal ledges, as in the Polaznensky karst region (see Sect. 3.4), are wave-break niches (Fig. 3.73a), caves, blocks of subsidence (Fig. 3.73b, c), pillar-like outliers.

A feature of the formation of leaching niches is an approximately twofold increase in their depth per year. Another feature is the confinement of the deepest niches to the shores not exceeding 10–15 m in height. With an increase in the coastal height, a decrease in the depth of the niches is observed. With a slope height of more than 30 m, niches with a depth of more than 5 m are rare. In contrast to the banks of reservoirs, composed of insoluble rocks, gypsum banks are characterized by high processing activity even in areas in the wave shadow. According to observational data, in the period from 1985 to 1998, the maximum rate of destruction of the coastal scarp was 4.2 m/year, and the average—0.3 m/year (Nazarov and Frolova 2009).

High dissolution rates are characteristic of gypsum that collapsed to the foot of the coastal scarp (Nazarov and Frolova 2009). Lumps of about 1 m^3 in size, being in the water, lost up to 70% of their volume in 2 years, turning into cellular fragments of an irregular shape, the so-called "Ural corals" (Fig. 3.73d).

A feature of the flow distribution in the karst strata is the presence of waterless pillars, the central part of which is practically not karst. In the underground (buried) relief, the roofs of sulfate layers of pillars are represented by outliers separated by depressions filled with karst-avalanche deposits. On the sides of depressions or on the slopes of pillars, karst cavities are recorded, which arise and develop, as a rule, under the influence of relatively low-mineralized hydrocarbonate-calcium waters of karst-avalanche sediments.

Karst massifs are composed of interbedded sulfate rock units and thin carbonate units separating them. In dolomites and limestones, karst is represented by secondary porosity and fine vugs, while, in anhydrite and gypsum, large cavities are formed along cracks. Dissolution and removal of sulfate rocks are accompanied by the subsidence of carbonate units with their simultaneous destruction and replacement by coarse-grained material with clay-rubble filler.

The karst content of sulfate rocks ranges from 1–5% in watersheds to 10–20% on the sides of river and dry valleys (Fig. 3.74). The vertical dimensions of the cavities range from fractions of a meter to 10 m or more.

Large caves often open on the slopes of the valleys. River waters are involved in their formation, invading the depth of the massif during high water and high floods. Underground galleries are formed along cracks in the northwest and northeast directions. At a distance from the river, narrow tunnels filled with clay can be traced.

Previously, six karst sites were identified within the region (Gorbunova et al. 1992).

The **Yurmano-Mulkovsky site** is dissected by the valleys of the Kutamysh and Yurman Rivers and ravines. Fissured gypsum and anhydrite 85–90 m thick are raised above river drains. Covered, subeluvial types prevail, and in the bend of the Sylva River near the Nasadka village, suballuvial types of karst prevail. Over an area of 34 km^2, more than 500 sinkholes and 9 depressions have been recorded; there are lakes, karst, and karst-erosional ravines. The average density of karst forms is 17, the maximum is 237 forms/km^2, and the coefficient of areal karst content is 0.32%. There are fresh dips on the site.

The **Serginsky site** is located mainly on the right bank of the Sylvinsky bay and includes the karst fields Yelkinskoye, Serginskoye, Dubrovinskoye, and Shatovskoye. The gypsum-anhydrite stratum has a thickness of up to 100 m and more (Fig. 3.75). It is overlain by Solikamsk terrigenous deposits, the thickness of which decreases from east to west until pinching out. Karst is covered, while bare and turfed along the banks of the Sylva and Serginka Rivers. On the

Fig. 3.73 Karst processes in the coastal zone of the reservoir (*photo by* N. Maksimovich). **a**—Wave-breaking niches in gypsum to the west of the Alebastrovo village; **b**—blocks of subsidence of gypsum overlain by carbonate rocks on the bank of the Sylva River to the west of the Alebastrovo village; **c**—blocks of gypsum and settling blocks on the bank of the Sylva River north of the Nasadka village; **d**—dissolution of blocks of gypsum that fell from the coastal ledge of the Sylva River, north of the Nasadka village

Fig. 3.74 Cheremukhovy dry karst valley north of the Plekhanovo village (*photo by* N. Maksimovich)

area of 63 km^2, 587 sinkholes, 3 kettles, and 7 ravines were studied, and 24 lakes were described. The average density of forms is 9.5, the maximum is 112 forms/km^2, and the average coefficient of areal karst content is 0.54%.

The **Nizhneshakvinsky site** is located on the right bank of the Shakva River. It is dominated by the carbonate-sulfate type of karst. On the area of 52.7 km^2, 492 sinkholes and subsidence, karst-erosion ravines, and 1 dry valley were

Fig. 3.75 Outcrops of gypsum on the right bank of the Sylva River to the west of the Serga village (*photo by* N. Maksimovich)

studied, and 20 lakes were registered. The average density of forms is 9.8, the maximum is up to 90 forms/km^2, and the coefficient of areal karst content is 0.5%.

The **Dubovsky site** occupies the most elevated part of the interfluve of Shakva and Sylva and is characterized by a carbonate-sulfate type of karst. There are no karst outcrops. On the area of 63.9 km^2, 855 sinkholes and subsidences, 1 depression (Sazhinskaya), 3 karst-erosion ravines, 1 dry valley, and 19 lakes were studied.

The karst lakes of the Sazhinskaya depression are up to 17 m deep and up to 150 m in diameter (Fig. 3.76).

In the **interfluve of Babka and Iren** Rivers, gypsum and anhydrite come to the surface in the eastern part along the banks of the Sylva River, in deep ravines and karst sinkholes (Fig. 3.77a). In the western part, they are overlain by Solikamsk rocks. Gypsum outcrops are observed in the steep cliffs of the Babka River valley (Fig. 3.77b).

The site is dominated by sulfate karst. The forms of its manifestation are funnels, karst-erosional ravines, cavities, and caves. The density of forms in some areas reaches 300 forms/km^2. Funnels with a diameter of 5–25 m prevail. Some of the funnels (87 out of 780) are filled with water or swampy. Numerous karst sinkholes in the eastern part of the interfluve are located in groups (near the Kochebakhtino and Bobry villages) or in chains in the direction from the Sogrino village to the Kaigorodovo village. In the vicinity of the Kochebakhtino village, in sinkholes, the diameter of which reaches 50 m, at the Sarkaevsky gypsum deposit of the Lunezhskaya suite, a closed, to an insignificant extent—bare and choked karst is developed. Numerous karrens, ponors, funnels, wells, and karst-erosion ravines are common on the steep bedrock slope of the Babka River and the adjacent territory, where in some places the coefficient of areal karst content exceeds 50%. The development of karst is facilitated by the cracks of the side resistance, to which small (5–10 m) crevice-like caves are confined. The coefficients of the areal and volumetric surface karst content of the entire deposit, respectively, are 10% and 50 cm. The gypsum top occurs at elevations from 135 to 188 m with the thickness of loose overburden deposits from 0 to 38 m.

On the territory of the Kungur town from 1948 to January 1, 2020, 697 failures were recorded. Long-term observations have shown that large-scale dips are a relatively rare phenomenon and arise in conditions of intense karst and waterlogged soils. The largest dip in the diameter of the depression and the depth of all those recorded in the territory of the Kungur town occurred on April 27, 2001, near the access roads to the correctional colony. The initial diameter of the sinkhole was about 10 m and the depth was 20 m. After some time, the diameter reached 21 m, and the sinkhole was filled with water coming from the surface and underground water from the bottom of the sinkhole. Around the edge of the sinkhole, concentric cracks in the soil were formed at a distance of up to 5 m. At the site of the sinkhole, the railroad tracks were deformed. The curvature of the rails was 22 cm vertically and 8 cm horizontally.

On the territory of the Nizhnesylvinsky karst region (excluding the territory of the Kungur town), 128 karst sinkholes were recorded over a hundred-year observation period. Many sinkholes are found in built-up areas,

Fig. 3.76 Karst lake in the Sazhino village (*photo by* N. Maksimovich)

Fig. 3.77 Outcrops of gypsum: **a**—in the valley of the Sylva River south of the Chuviryata village; **b**—on the right bank of the Babka River near the Yergach village (*photo by* N. Maksimovich)

Fig. 3.78 Sinkhole near the Rodionovo village, 2018 (*photo by* O. Kadebskaya)

Fig. 3.79 Karst sinkhole in a field near the Lukino village, 2019 (*photo by* O. Kadebskaya)

highways and railways. Common sinkhole sizes are up to 3 m in diameter and up to 4 m in depth. Nine dips had a maximum size of 8–14 m in diameter and a depth of 5–18 m. The largest one occurred in 1985 in the Rodionovo village, the initial diameter was 20 m at a depth of 15 m. At the time of its formation, the hole opened up an underground cavity 120 m long, which is currently inaccessible. In 2018, in the vicinity of the Rodionovo village, another sinkhole occurred, 10 m in diameter and 18 m deep (Fig. 3.78). In 2019, near the Lukino village, when plowing farmland after a tractor had passed, a failure occurred with dimensions of 18 by 16 m and a depth of 10 m (Fig. 3.79).

The confinement of most of the large dips to Pliocene terraces is explained by the significant height of their basement and the density of the overburden. Large horizontal galleries are formed in sulfate sediments under the cover of carbonate rock units. In this regard, the presence of such rocks above sulfate rocks may indicate their increased karst content and the presence of large voids (Lukin and Ezhov 1980).

In the cover sediments of young and ancient terraces, directly overlying karst rocks, two horizons are distinguished: the upper—alluvial deposits, and the lower—karst breccias, represented by blocks, gruss, for the most part, carbonate rocks with inclusions of sandy-argillaceous and gravel-pebble material. In the lower part, the amount of sulfate rock fragments increases. Karst breccias occur on intensely karst sediments. The horizon of karst breccias is characterized by an inhomogeneous thickness from 1–2 to 20–30 m and occurs at the base of the blanket sediments throughout the entire area of gypsum anhydrite deposits (Lavrova 2005).

Caves. In the archives of the Kungur laboratory of the Mining Institute of the Ural Branch of the Russian Academy

Table 3.4 The largest caves in the Nizhnesylvinsky karst region (Caves of the Volga … 2010; with additions by the authors)

Cave name	Morphometric indicators, m		
	Length	Amplitude	Depth
Sylvinsky speleo-subregion			
Kungur Ice Cave	5700	30	12
Zuyatskaya Cave	1410	30	30
Scouts (Novaya Podkamenskaya) Cave	1200	15	–
Kichmenskaya Ice Cave	470	33	33
Monastyrskaya 1 Cave	410	5	3
Bolshaya Mechkinskaya Cave	465	25	25
Zakurinskaya Cave	300	25	22
Oktyabrskaya 1 Cave	290	9	8
Oktyabrskaya 2 Cave	194	13	11
Malaya Kungurskaya	150	20	19
Podkamenskaya (Bolshaya Podkamenskaya) Cave	150	5	5
Alabastrovskaya Cave	150	–	–
Medvezhyegorskaya (Nastenkina) Cave	115	17	17
Monastyrskaya 2 Cave	100	2	–
Pike Mouth (Alabastrovskaya 2) Cave	95	–	–
Podkamenskaya 2 Cave	60	–	–
Dyrikhinskaya Cave	60	–	–
Cemetery 1 Cave	9	–	–
Babkinsky speleo-subregion			
Babinogorskaya Cave	950	26	26
Srednaya Kazaevskaya Cave	190	6	6
Nignaya Kazaevskaya Cave	80	–	–
Yergachinskaya Cave	68	8	–

(*Note* "–"—no data)

of Sciences and the Association of Speleologists of the Urals in the Nizhnesylvinsky karst region, there were 86 caves and grottoes. The largest are the Kungur Ice Cave (5700 m), Zuyatskaya (1410 m), Skautov (1500 m), Babinogorskaya (950 m), and Bolshaya Mechkinskaya (465 m) caves (Table 3.4).

The **Kungur Ice Cave** is the largest gypsum cave in the Cis-Urals, its length is 5700 m. It is a unique natural site of world importance. It is located on the right bank of the Sylva River, in the southwestern part of Ledyanaya Mountain. The entrance to the cave is located at the base of the rocky outcrop (Fig. 3.80).

The cave is widely known in Russia and abroad as one of the most interesting and visited sites of speleotourism. Since the beginning of the twentieth century, more than 5 million people have visited the cave. Annual attendance in recent years has reached an average of 100–130 thousand people.

During 300 years of research, about 100 popular scientific papers and over 500 scientific works have been devoted

Fig. 3.80 Outcrops of gypsum and dolomites that make up the Ledyanaya Mountain on the right bank of the Sylva River (*photo by* O. Kadebskaya)

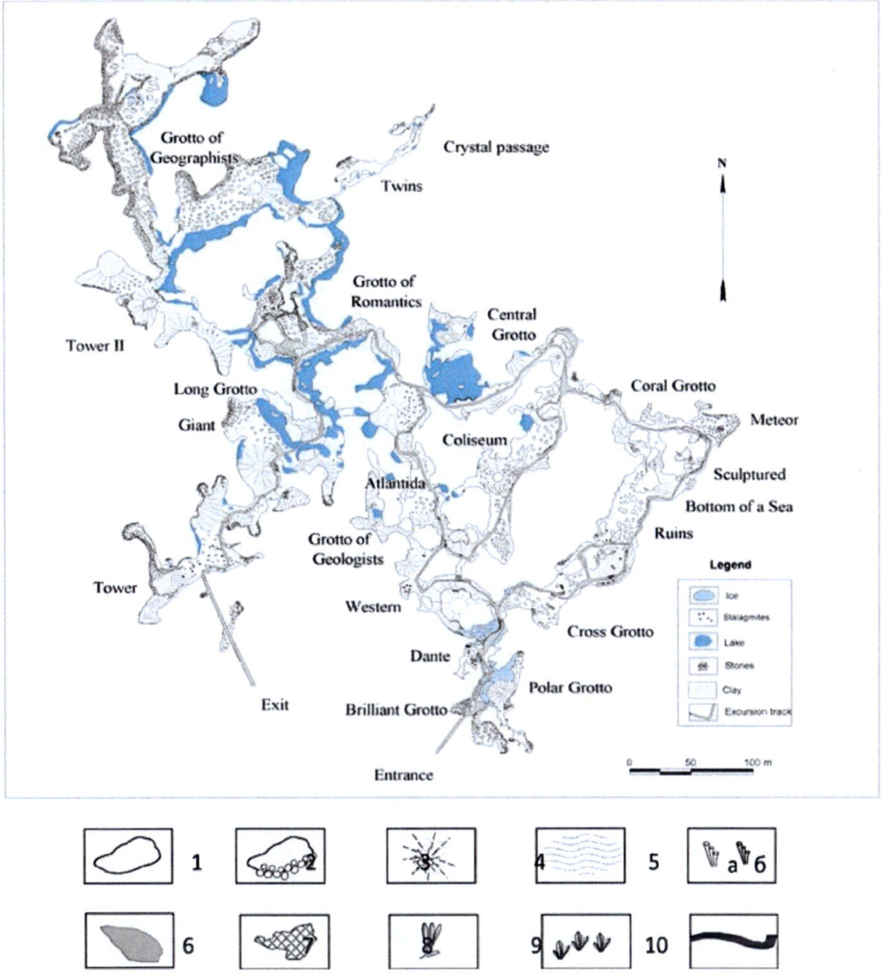

Fig. 3.81 Plan of the Kungur Ice Cave (Dorofeev 1967). 1—Walls made of bedrock; 2—walls made of destroyed rocks; 3—blocky-gravelly-clayey deposits; 4—clay; 5—organ pipes: **a**—gaping, **b**—filled; 6—lakes, 7—cover ice, 8—stalagmites, 9—crystals, 10—excursion trail

to the cave (Gorbunova et al. 1997, 1998; Gorbunova and Maximovich 1994a, b). Over the past 50 years, detailed stationary geological, speleological, biological, and other observations have been carried out there. The peculiarity of the cave is the predominance of large grottoes, which appear at the intersection points of the galleries and when they are dissected by rubble (Fig. 3.81). The cave is a horizontal labyrinth stretching from the side of the Sylva River valley (at the level of its first above-floodplain terrace) into the depths of the Ledyanaya Mountain massif for 700 m.

In many grottoes of the cave, there are numerous (about 70) underground lakes with a total area of about 7.5 thousand m^2 (11% of the total area of the cave). In the entrance part of the cave, there is perennial ice and, during the winter period, a large number of ice formations are formed (Fig. 3.82). Detailed information about the cave is given in the monograph (Kungur Ice Cave … 2005).

Fig. 3.82 Krestovy grotto, Kungur Ice Cave (*photo by* O. Kadebskaya)

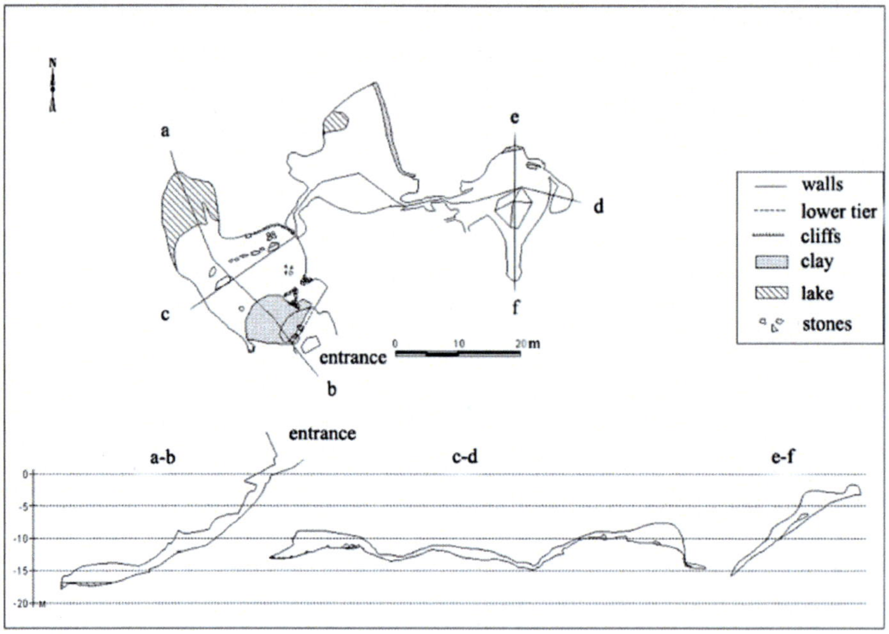

Fig. 3.83 Plan of Malaya Kungurskaya Cave (Lavrov, Zemlyanikin, Zubanova, Sergeeva 2000)

The **Malaya Kungurskaya Cave** is located on the right bank of the Sylva River, 700 m below the Kungur Ice Cave, not far from the Medvezhiy Log. The entrance to the cave is located on the side of a sinkhole on the slope of Ledyanaya Mountain, 20 m above the water's edge. There was an ice plug at the bottom of the funnel. In the summer of 1999, the entrance to the first grotto was discovered behind a 1.5 m layer of ice. In the winter of 2000, speleologists from Chelyabinsk under the leadership of P. Sivinskikh, after clearing the passage, discovered another grotto. In the spring of the same year, cavers under the leadership of I. Lavrov dug an 8 m passage in the western wall of the second grotto and entered the third (Fig. 3.83).

There are two lakes in the cave, in the spring in the second and third grottoes a stream flows periodically during floods. In the first and second grottoes, various ice formations are observed in winter. A small amount of perennial ice was noted at the entrance. Due to the runoff of water in the spring, the entrance is covered with ice, and the cave is closed for several years.

The **Zuyatskaya Cave** is located in Sukhoy Log, which is 1 km northwest of the Zuyata village (Fig. 3.84).

The cave has long been known to the locals. In 1952–1955, it was described by the staff of the Kungur stationary under the leadership of A. V. Turyshev, in 1959, it was investigated by the expedition of the Perm University (Bobrova 1960), and, in 1983, speleologists from the Perm city drew up its new plan (Valuisky et al. 1988).

The entrance to the cave, measuring 5 by 1.5 m, is located at the foot of a small rocky outcrop on the right slope of the ravine. The cave is an extensive system of passages and grottoes with streams and lakes with a total length of 1410 m and the volume of grottoes is not inferior to the Kungur Ice Cave (Fig. 3.85). A special feature of the cave are five vast lakes with clear water, one of which has a long underwater gallery. Lake levels are subject to significant seasonal fluctuations. In one of the lateral branches, a stream flows, forming small waterfalls.

Fig. 3.84 The sinkhole over the Zuyatskaya Cave (*photo by* O. Kadebskaya)

The **Bolshaya Mechkinskaya Cave** is located 20 km north of the Kungur town and 2 km from the settlements (Zaspalovo and Rodionovo), on the left bank of the Mechka River. The entrance is located in the estuarine part of the right slope of the Kamenny Log, at the base of a rock outcrop of gypsum and anhydrite of the Kungurian stage. The cave consists of five grottoes and passages connecting them with a total length of 465 m (Fig. 3.86). In winter, the cave is rich in

Fig. 3.85 Lake in the Zuyatskaya Cave (*photo by* Y. Zanda)

ice crystals, stalactites, stalagmites, and columns. Three invertebrate species were found here: nematodes, amphipods, and winter mosquitoes (Geological monuments … 2009).

The **Zakuryinskaya Cave** is located on the right bank of the Sylvinsky bay of the Kama reservoir, 0.5 m east of the Zakurya village. The arched entrance to the cave, 3.5 m wide and 3 m high, is located at the base of a steep rocky slope of a large karst sinkhole, 30 m from the coastal cliff.

From the entrance grotto with dimensions of 7 by 15 m and an average height of 2 m, two well-shaped passages depart, which lead to the main grotto with a lake (Fig. 3.87). Its length is about 30 m, width is up to 15 m and depth is up to 9 m. The water level in the lake varies depending on the water level in the reservoir. In the 1950s, before filling the

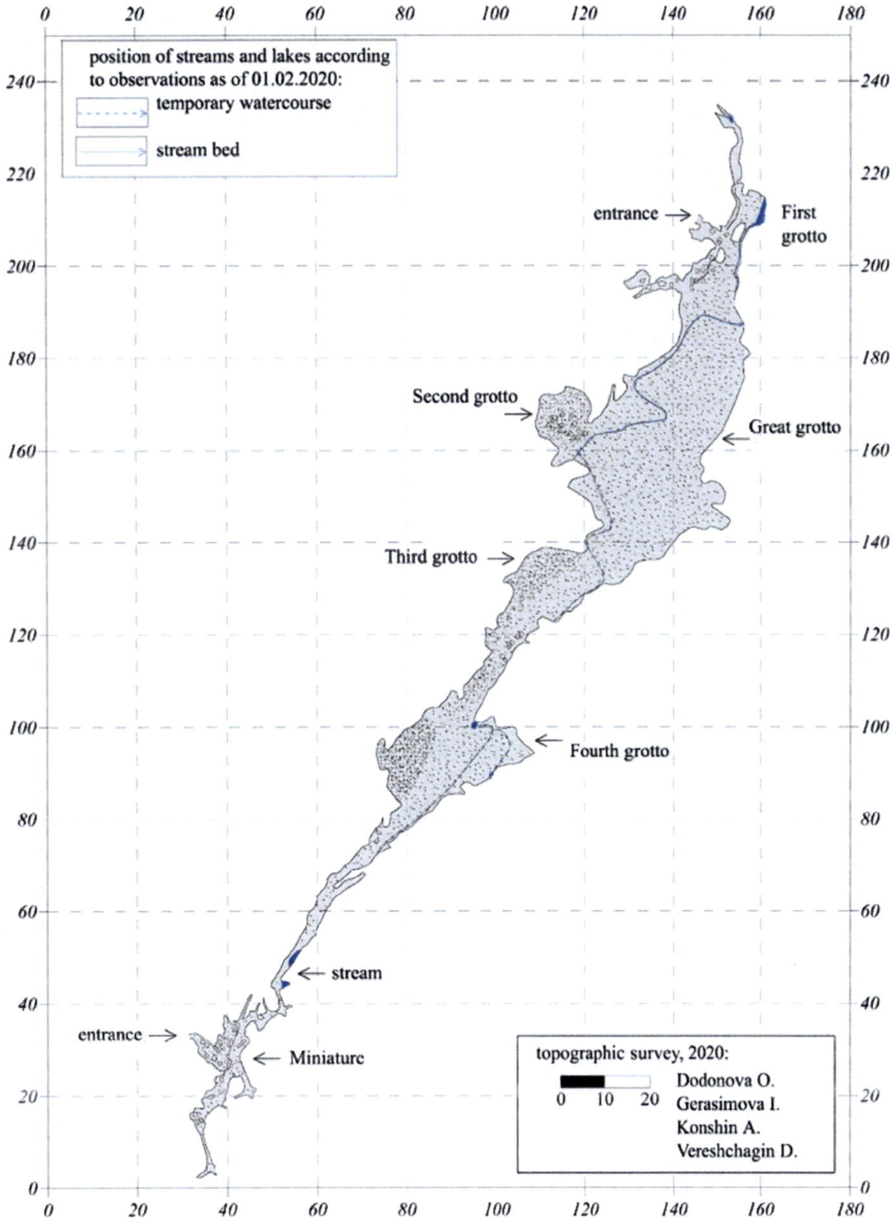

Fig. 3.86 Plan of Bolshaya Mechkinskaya Cave (Dodonova, Gerasimova, Konshin, Vereshchagin 2020)

Fig. 3.87 Lake in the Zakurinskaya Cave (*photo by* Y. Zanda)

Fig. 3.88 Karst Lake Nuzhino (*photo by* O. Kadebskaya)

reservoir, the cave was investigated by A. V. Turyshev, an employee of the Kungur stationary, who plotted 300 m of passages on the plan of the cave. In 2018, the entrance part of the cave collapsed, as a result of which the main passage was filled with surface sediments, clay, and sandy loam. Currently, the length of the accessible part does not exceed 78 m.

Existing and proposed specially protected natural areas

Existing SPNAs. Within the Nizhnesylvinsky karst region, there are 11 SPNAs of regional and local significance associated with karst phenomena, with a total area of 2,075.66 ha with a buffer zone of 53.9 ha.

Baydarashki is a geological natural reserve of regional importance with an area of 125 ha. It is located between the Sylva and Shakva Rivers, on the northwestern edge of Ledyanaya Mountain. The geological section is represented by the carbonate-sulfate strata of the Kungurian stage. The main feature is the maximum intensity of karst sinkholes for the Perm Region: up to 1 thousand/km^2, a bright example of a karst landscape.

The Bolshaya Mechkinskaya Cave is a geological natural monument of regional significance with an area of 6 ha and a protected zone of 23.1 ha. It is located 20 km from the Kungur town.

The Zakurinskaya Cave is a geological natural monument of regional significance with an area of 3 ha and a protective zone of 9.1 ha. Located in the vicinity of the Zakurya village. The total length of all passages in the cave is 300 m.

The Zuyatskaya Cave with an area of 10 ha and a protective zone of 2.8 ha has the same status. It is located 1–1.5 km northwest of the Zuyata village. It ranks fifth among the largest in the Perm Region. Landfall deposits are developed in the form of fragments of gypsum and clay residues.

The Ledyanaya Mountain and the Kungur Ice Cave are a historical and natural complex of regional significance with an area of 106 ha. It is located on the territory of the Filippovka village, Kungur region, and the Kungur town. It is the largest gypsum cave in the Perm Region.

Lake Nuzhino is a natural reserve of local importance with an area of 0.36 ha on the territory of the Zaborino rural settlement of the Berezovka district. It is located in a funnel on the banks of the Shakva River, at the bottom of which several large karst springs are unloaded. A stream flows out of the lake and flows into the Shakva River (Fig. 3.88).

Lake Krotovskoye is a geological monument of local importance with an area of 4.6 ha and a conservation zone of 2.1 ha. It is located on the territory of the Kungur town. It is a karst sinkhole filled with water.

Lake Povarennoe with an area of 78 ha is located in the territory of the Kungur town. It is a filled karst funnel, the water from which goes underground with a periodicity of 10–15 years.

The **Perm-Serginskaya karst rocky steppe** is a landscape natural monument of regional significance with an area of 2.7 ha and a conservation zone of 16.8 ha. It is located on the northern outskirts of the Serga village and occupies the right root bank of the Sylvinsky bay of the Kama reservoir. It is an area of sulfate karst with the northernmost area of feather-grass forest-steppe and the Andronovsky grotto with a lake (Fig. 3.89).

Spasskaya and Podkamennaya mountains are a protected landscape of regional significance with an area of 1,740 ha. It stretches for 15 km from the Plekhanovo village to the Kashirino village. It occupies a part of the right elevated bank of the Sylva River, where powerful intensive karst outcrops of gypsum and anhydrite of the Irenian horizon stretch for several kilometers. Karst forms are widely developed: ravines, ditches, ledges, rocks, and outliers (Figs. 3.90 and 3.91).

Fig. 3.89 Entrance to the Andronovsky grotto (*photo by* Y. Zanda)

Fig. 3.90 Outcrops of gypsum on the right bank of the Sylva River near the Kashirino village (*photo by* N. Maksimovich)

Several large caves are known in Podkamennaya Mountain. The main process leading to the formation of caves here is the gravitational process—unloading of slopes. This type of karst cave is oriented along the slopes of river valleys and ravines, and their formation is associated with the dissolution of gypsum and anhydrite along discharge cracks.

Proposed SPNAs. Kichmenskaya Cave until 2009 was a geological natural monument of regional significance. According to the authors, it was unreasonably excluded from the protected areas. It is located on the right bank of the Kichmen River, 4.5 km above the confluence with the Yurman River. The entrance to the cave is located in a karst gap 11.5 m deep. It is necessary to restore the protected status of the cave, in addition, to submit for consideration a unique section of the **Yelkinsky outcrop**, which is a stratotype of the Yelkinskaya suite of the Irenian horizon of the Lower Permian (Fig. 3.71). The section is located on the right bank of the Sylva River below the Yelkino village. Hypsoanhydrite strata are devoid of organic remains and differ only in their position relative to carbonate layers, which play the role of marking horizons (Geological monuments … 2009).

Economic activity and environmental pressure. Ozhginskoye, Ilyichevskoye, and Yurmanskoye oil and gas fields are being developed within the region, gypsum is being mined, and agriculture is developed. Sections of the Sverdlovsk railway, the Perm—Yekaterinburg, Perm—Chusovoy highways pass through the district. In the north of the region, logging is carried out in the Yurman River basin.

There are large **deposits of gypsum and anhydrite** in the region.

The Shubinskoe gypsum deposit is located on the left bank of the Iren River, 3 km from the Shubino village. There are karst sinkholes within the allotment. During the development of gypsum, karst cavities were uncovered, partially filled with grit-rubble soil, limestone blocks, dolomites, marls, and gypsum mixed with loams and clays. The deposit is not watered, belongs to the third group of geological complexity due to the sharp variability of the thickness and quality of the mineral (Lavrova 2013).

The Deikovskoe gypsum deposit is located on the left bank of the Sylva River, 4 km from the Mokhovoe village. During the study of core samples from exploration wells in the field, a rare borate mineral, probertite, was described in gypsum (Galimov et al. 2013).

The Yergachinskoye gypsum deposit is located in the Kungur district, 3 km from the Yergach railway station. The development of the field is at the final stage of reclamation.

The Sokolino-Sarkaevsky gypsum and anhydrite deposit is located 2.5 km from the Yergach railway station, nearby is the Yergach gypsum plant, which processes gypsum and anhydrite. The deposit has been developed since 1928. Sulfate-carbonate rocks of the Lunezhskaya suite of the

Fig. 3.91 Podkamennaya Mountain, view from the side of Nogayev log (*photo by* O. Kadebskaya)

Irenian horizon of the Kungurian stage of the Lower Permian take part in the geological structure of the useful strata. The gypsum unit is heavily karst. Karst forms are manifested both on its surface—buried ancient karst funnels, opened during the exposure of the gypsum roof, and inside it—filled and unfilled karst cavities, opened by wells. The anhydrites are not karsted (Khudenkikh and Kataev 2018).

The Razepinskoe gypsum and anhydrite deposit is located 1 km from the Shavkunovo village, Kungur region. The gypsum stratum is mined by an open-cut method, and the thickness of the overburden is 30 m. According to the calculation of the reserves in 2012, the mineral resources will be sufficient for 47 years.

On the territory of the district, there are enterprises that are engaged in crop production, animal husbandry, and processing of agricultural products. In the areas of farms, poultry farms, etc., nitrogen-type waters with a predominance of nitrate ions up to 535 mg/dm^3 were registered (Minkevich and Kilin 2018), which increases the solubility of gypsum. The high karst content of the region contributes to the spread of these waters. Groundwater, rivers, and lakes within settlements used by the local population for water supply are polluted.

Railway transport. A section of the Sverdlovsk railway with a length of 90 km passes through the territory of the district. On the section, 1416–1518 km of the Yergach–Iren section in 1989, there was a sinkhole under the locomotive of a passenger train (Gorbunova et al. 1992). Subsequently, a detailed field survey of karst forms was carried out on the operating hauls using the radio wave method. More than a dozen large and small karst forms have been identified and delineated (Zaderigolova et al. 2015).

In 1993, on the Yergach–Iren stretch, in the 1516 km section, a sinkhole with a diameter of 24 m and an apparent depth of up to 5 m occurred again (Fig. 3.92; Golikov 2017).

In 1897, a bridge across the Sylva River was built on the Ural Mining Railway (now Sverdlovskaya) near the Sylva village (Fig. 3.93). At a depth of 22–25 m on the floodplain and 11–12 m under the channel sediments, gypsum and anhydrite of the Irenian suite of the Kungurian stage occur. Gypsum is fissured, and cavernous in places. A dense bluish-gray anhydrite occurs below a depth of 43–53 m. After the construction of the bridge, uneven settlement of the support and its cracking were found. Observations from 1931 to 1947 showed an increase in its slope, and the crack width increased from 65 to 97 mm. To establish the causes of this phenomenon, a well was drilled, which at a depth of 24.7 m opened a cavity with a vertical size of 0.8 m. In other

Fig. 3.92 Sinkhole on the section of the 1516th km of the Sverdlovsk railway, the Yergach–Iren section, 1993 (Golikov 2017)

wells, cavities filled and not filled with debris with a diameter of up to 6.1 m were also registered (Maksimovich and Gorbunova 1958).

Automobile transport. There is a well-developed road network in the region. A major dip occurred on November 17, 2017, in the Perm Region on the Mostovaya–Syra road leading to a ferry crossing to the Nasadka village (Fig. 3.94). The hole with a diameter of 10 m and a depth of 13 m was formed during repair work.

On May 7, 2020, a sinkhole is formed at the edge of the asphalt pavement of the Kungur–Serga road, 150 m north of the Zaspalovo village (Fig. 3.95a). The shape of the dip in plan is oval, the maximum diameter is 2 m, the minimum is 1 m, and the apparent depth is 0.6 m. In this area higher up the slope in 2013–2014, a number of small dips were recorded at the place where the surface runoff was absorbed by the drainage ditch (Fig. 3.95b).

The sinkhole near the highway, 2.6 km south of the Durakovo village, Kungur district, was first surveyed on June 1, 2009. The shape was round in plan and pear-shaped in cross-section. The cross-section on the surface was 2 m, increasing with a depth of up to 10 m. At present (summer 2020), the sinkhole has a diameter of 16 m and a depth of 9.3 m (Fig. 3.96).

3.7 Kishertsky Region of Predominantly Sulfate and Carbonate-Sulfate Karst

The region stretches in a narrow strip with an area of 334 km^2 along the eastern slope of the Ufimian plateau from the interfluve of Sylva and Shakva Rivers to the southern border of the Perm Region (Fig. 3.97). Administratively, it is located within the Berezovka, Kishert, Kungur, Suksun districts

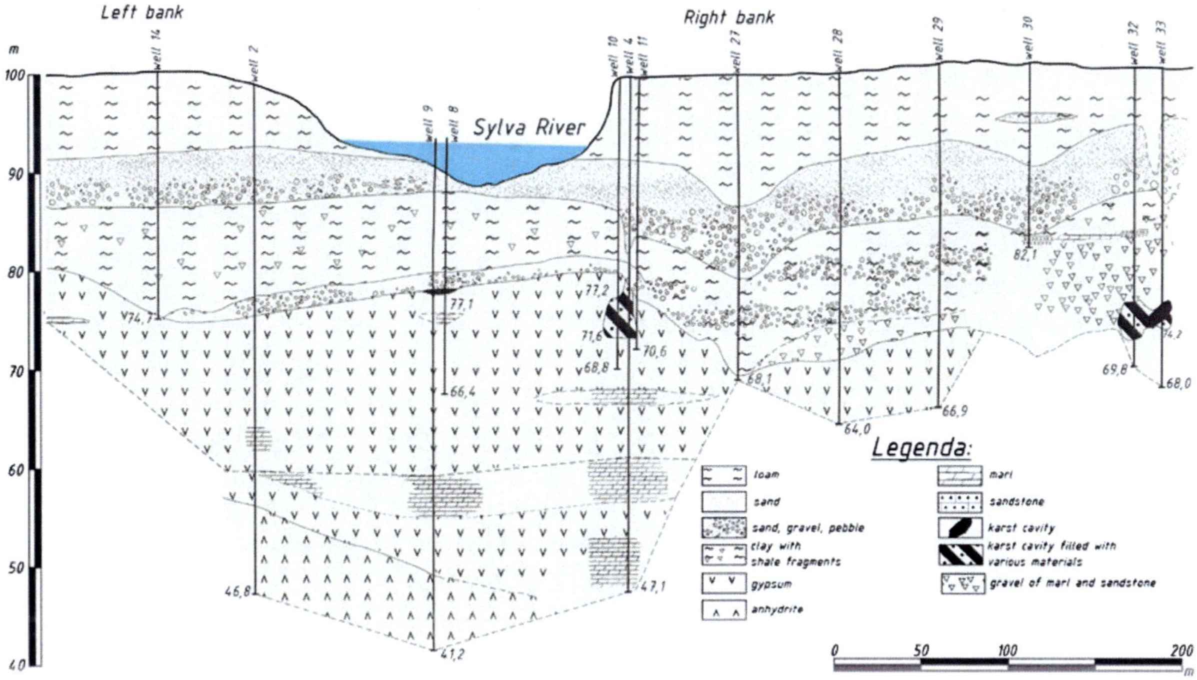

Fig. 3.93 Geological profile of the bridge over the Sylva River (archived data of the laboratory of the geology of technogenic processes of the Natural Science Institute, Perm State National Research University)

Fig. 3.94 Sinkhole on the road Mostovaya–Syra (*photo by* I. Kostarev)

of the Perm Region and occupies mainly the left-bank part of the middle course of the Sylva River, which crosses it in the north in a latitudinal direction (Gorbunova et al. 1992).

Geological conditions for the development of karst. The region is located in a narrow junction zone between the East European Platform and the Ural Foredeep and is confined to the Yuryuzan-Sylva depression. The Kishertsky karst region, according to Gorbunova (1992), is the key one in the study of the sulfate karst of the Perm Region (Fig. 3.98).

Geologically, it corresponds to a narrow contact zone of carbonates of the Ufimian swell (the eastern margin of the East European platform) and terrigenous rocks of the western margin of the Cis-Ural trough with interlayers and lenses of limestone, gypsum, and rock salt of the Leksky, Popovsky, and Koshelevsky formations of the Philippovsky and Irensky horizons of the Kungurian stage. Carbonate deposits of the Artinskian and Philippovsky horizons of the Kungurian stage on the eastern flank of the Ufimian swell subside towards the trough and are replaced by clays, siltstones, marls, clayey limestones, and anhydrites of the Leksky formation. As they sink, carbonate deposits are overlain by the Popovsky formation (Irenian horizon) of the Kungurian stage, represented by marls, dolomitized marls, clays, clayey limestones, and sandstones with gypsum and anhydrite interlayers and lenses. At its base, a layer of anhydrites can be traced. Gypsum and anhydrite of the Popovsky formation come to the surface near the Ust-Kishert village. Partly, in the section, rockfall-karst cemented sediments are contained, the thickness of which reaches 45 m. On the watersheds, in ancient karst depressions, Neogene white, and colored clays, quartz sands, and pebbles can be met, less often—boulders and gravel of cherts, drain pebbles, and quartzites. The Quaternary system is represented by alluvium in river valleys, eluvial-diluvial, proluvial, and lacustrine-bog sediments.

Hydrogeological conditions. The Kishert-Irginsk water-abundant zone is distinguished within the Kishertsky karst region.

a *b*

Fig. 3.95 A small dip (**a**) and an opened organ pipe in a ditch (**b**) of the Kungur–Serga road north of the Zaspalovo village, 2020 (*photo by* N. Maksimovich)

Fig. 3.96 A sinkhole 30 m from the Kungur–Serga road near the Durakovo village, 2020 (*photo by* N. Maksimovich)

The Sylva River is the main watercourse of the region and belongs to the type of river with mixed feeding, the share of underground feeding can be 15%. Its basin lies in a zone of moderate continental climate and is characterized by excessive moisture. From 650 (southern part of the basin) to 800 mm (northern part) of precipitation fall annually. The river belongs to the Eastern European type with a pronounced spring flood, summer-autumn rain floods, and a long stable winter low-water period. Mixed feeding with a predominance of snow and rain is characteristic: up to 60% of precipitation flows into rivers, and the distribution of river flow within a year is relatively uniform. The water discharge in the channel fluctuates within significant limits: from a minimum of 53.1 m^3/s in March to a pronounced maximum of 536.5 m^3/s in May; the average annual discharge is 1804.5 m^3/s (Mikhailov and Oborin 2006).

The water of the Sylva River has a constant calcium-hydrocarbonate-sulfate composition, and the salinity of the water increases from the spring (475.0 mg/dm^3) to the winter (717.0 mg/dm^3) period.

The region is a discharge zone for karst waters of the Ufimian swell. Waters are localized along fissure and karst zones, and lithological contacts. Such places of groundwater discharge on the eastern flank of the Ufimian swell include the source of the Kishertka River, the discharge of water at the mouth during low-water periods is about 60 l/s, and during floods—up to 200 l/s (Fig. 3.99a).

The chemical composition of the waters of the Kishertka River changes downstream, has seasonal character, and depends on the feeding conditions. In the upper reaches near the Nizkoye village, river water has a low mineralization (497–535 mg/dm^3) and a calcium-magnesium-hydrocarbonate-sulfate composition due to dilution with fresher waters coming from carbonate sediments (Fig. 3.99b). Mineralization increases near the mouth and varies from 1034 to 1408 mg/dm^3 during the year, which is associated with the discharge of sulfate waters in this part of the region. The predominant ions in the spring, autumn, and winter periods are Ca^{2+}, SO_4^{2-}, and HCO_3^-, while in the summer, the composition transforms, and the waters become calcium-magnesium-sulfate-hydrocarbonate.

In the interfluve of Sylva and Kishertka Rivers, there is a watershed of groundwater (according to the scheme of Turyshev 1960), from which water moves in the northern and northeastern directions in relation to the rivers. In the area of the Ust-Kishert village, on a relatively small area, all types of discharge of karst waters are observed, which are recorded by changes in the composition and mineralization of water manifestations: ground (spring), subaqueous in river beds, and at the bottom of lakes, underground in waters

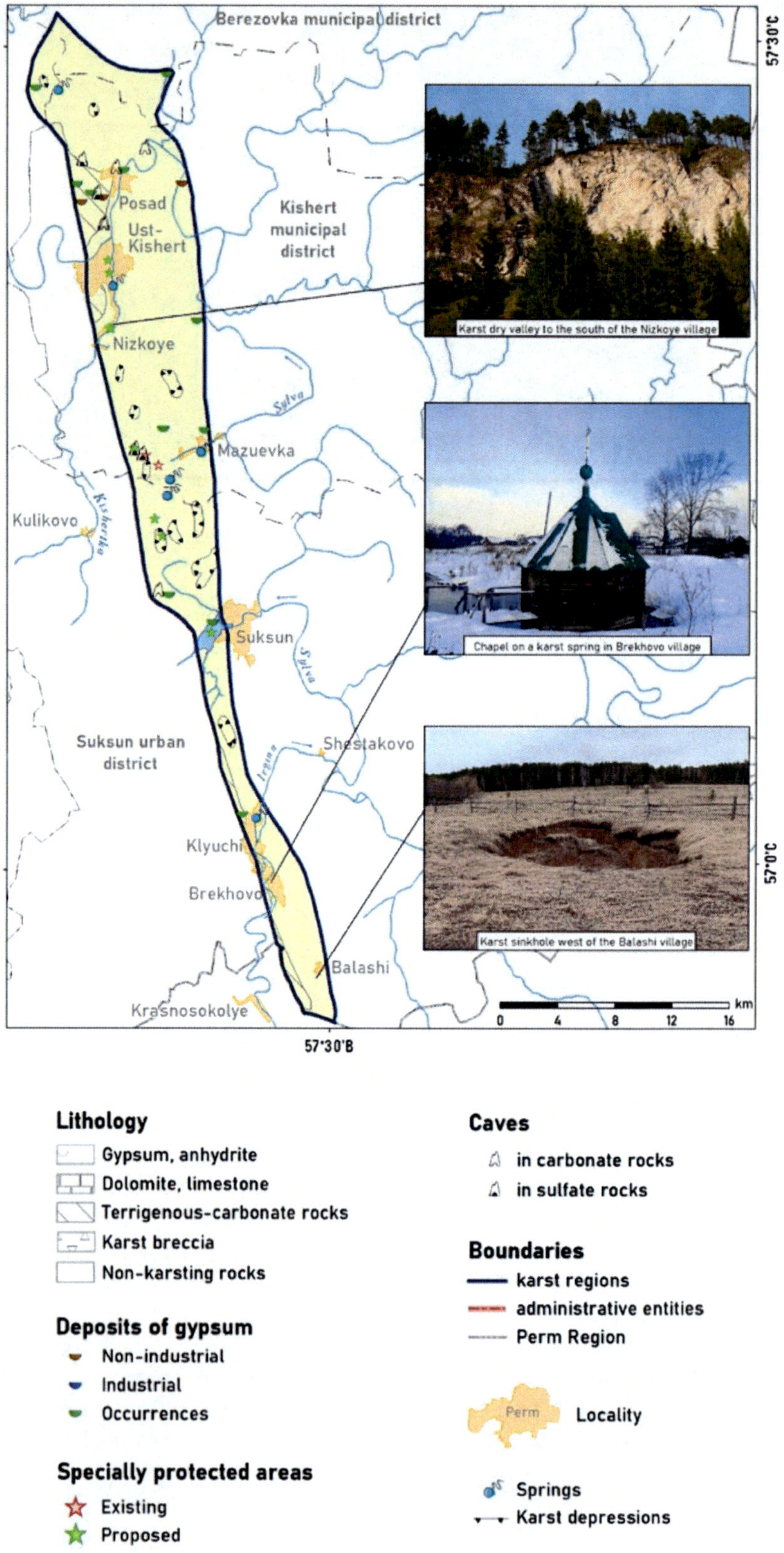

Fig. 3.97 Kishertsky karst region

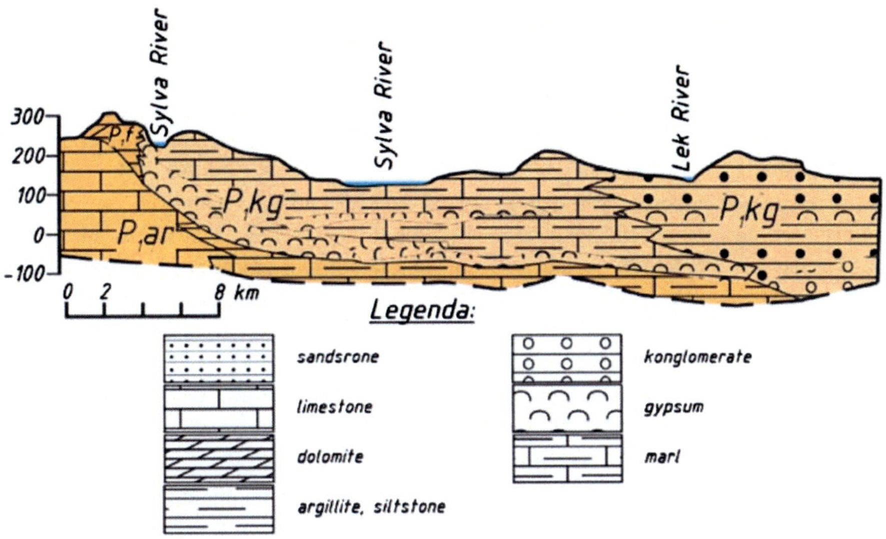

Fig. 3.98 Geological conditions for the development of karst in the Kishertsky region (Gorbunova et al. 1992)

Fig. 3.99 Kishertka River (*photo by* N. Maksimovich). **a** Near the source next to the Kovaleva village (Mikhanyata); **b** above the Nizkoye village (Sharakhino tract); **c** dry valley within the Nizkovsk depression; **d** the place where it flows into the Sylva River in the Ust-Kishert village

of other aquifers in contact with gypsum and anhydrite. The underground waters of alluvial deposits in the valleys of the Sylva and Kishertka Rivers, as well as fissure-karst waters of the gypsum-dolomite-marl strata, are hydraulically interconnected (Fig. 3.99d).

An intensive manifestation of karst processes is associated with the inflow of fresh hydrocarbonate waters from the side of the Ufimian swell and their movement along the most disturbed fractured zones or lithological contacts into gypsum-anhydrite deposits.

The aquifer of carbonate deposits is divided here into concentrated streams directed to the discharge sites along the most fractured and karst zones, as well as along the contacts of dissimilar rocks. Streams of karst waters seem to bend around reefs, gypsum and anhydrite massifs, less fractured and low-permeability rocks, and the runoff is not only transverse but also longitudinal relative to the shaft axis. The waters of the zones of horizontal and siphon circulation of the Ufimian swell are discharged in the form of descending and ascending springs, as well as by filtration into alluvium, karst breccias, fractured weathered sulfate deposits. At the contact with sulfate deposits, fresh hydrocarbonate-calcium waters are transformed into sulfate–calcium waters with a salinity of up to 2.6 mg/dm^3 (Gorbunova et al. 1992).

In the section of the zone between the Kishert and Klyuchi villages, the total flow rate of the springs coming out from the side of the platform is 1109 l/s. Groundwater outflows to the surface are concentrated in three places: on the left bank of the Sylva River below the Mizuevo village (more than 25 l/s), at the mouth of the Suksun dry valley (541 l/s), and in the valley of the Irgina River (543 l/s).

Budanov (1964) notes that the flow rate of large springs in the characterized zone of intense fracture disturbances reaches 1000 l/s, and the specific flow rate of wells is 135 l/s (the Klyuchi village).

In the Nizkovsky site, the total flow rate of groundwater discharges during the period of dominance of underground recharge is 60 l/s. The rivers flowing in the northern part of the zone have the following flow rates: the Kishertka River on the northern outskirts of the Kishert village—160 l/s, the Sedinka River in the mouth part—50 l/s, and the Sovyanka River—120 l/s.

The total spring discharge of karst waters, both hydrocarbonate and sulfate in the Kishert region, is more than 2500 l/s.

The chemical analysis of the sources of the Ust-Kishert village showed that the underground waters discharging near the contact of carbonate-sulfate and terrigenous rocks are characterized by a calcium-magnesium-hydrocarbonate-sulfate (source of the Nizkoye village) and calcium-hydrocarbonate (Lakes Provalnoe and Molebnoe) composition and mineralization of 144–738 mg/dm^3. In spring, the waters of Lake Provalnoe contain magnesium ions associated with the dissolution of dolomite. Molebnoe and Provalnoe are lakes of mixed, predominantly underground, and water supply of cover and bedrock sediments (Kataev et al. 2017). They are characterized by vertical hydrochemical zoning associated with changes in chemical parameters and water type with depth. The change in the composition of the waters of Lake Provalnoe from calcium-hydrocarbonate to calcium-hydrocarbonate-sulfate and a sharp increase in mineralization indicates an inflow of groundwater from the sediments of the Kungurian stage. The increase in the mineralization of the waters of Lake Molebnoe is associated with an increase in the content of calcium ions and hydrocarbonate ions, and not in the amount of sulfates. Groundwater inflow is carried out from carbonate deposits (Maksimov et al. 2017).

The spring in the Zuevo village (Fig. 3.100), which emerges within the distribution of terrigenous rocks of the Kungurian stage, has a calcium-sulfate composition and high mineralization (1623.0–2110.0 mg/dm^3). This is due to the fact that the replacement of carbonate-sulfate terrigenous rocks of the trough from west to east creates conditions for the pressure discharge of fresh hydrocarbonate waters, accompanied by mixing with sulfate waters (Kazantseva et al. 2020).

The main drain in the southern part of the Kishert-Irginskaya zone is the Irgina river, whose discharge

Fig. 3.100 Unloading of sulfate–calcium waters in the form of a source in the village of Zuevo (*photo by* N. Maksimovich)

in the mouth part near the Shestakovo village in July 1968 was 5.8 m^3/s. During the summer low-water period, underground food predominates here. The inflow of groundwater in the area between the Klyuchi village and the dam in the Nizhniy Irginsk village is 4.5 m^3/s. The largest inflow of groundwater was established in the sections of the Klyuchi village—the Brekhovo village (1930 l/s), the Krasnosokolye village—the Pantino village (900 l/s), and the Brekhovo village—the Shchipitsino village (880 l/s).

Water-abundant areas in the Irgina River valley alternate with non-water-abundant ones, where the inflow of groundwater is small or absent. In some areas, channel water absorption occurs. The specific inflow of groundwater into the channel of the Irgina River reaches 480 l/(s km^2) and is on average 118 l/(s km^2).

The Soyanka River, 7.6 km long, flows into the Suksunsky pond (Fig. 3.101). It is started by springs with a small debit flowing from the marls of the priest's suite near the Chechuvarki village. The discharge of the river from the source downstream slowly increases from 0.7 to 18 l/s (4.8 km from the mouth). In the middle course on the left, the Klyuchevaya Kamenka River flows into it with a flow rate of 4 l/s. Despite this, the flow rate is reduced to 16 l/s (2.8 km from the mouth). According to local residents, in dry years the river disappears in this area. In the lower course, numerous ascending springs with a flow rate of up to 54 l/s flow into the Soyanka River on the left, below which the flow rate increases, and at the mouth is 257 l/s. The chemical composition of the river is determined by the feeding conditions. In the upper and middle reaches, the water has a salinity of 486–507 mg/dm^3 and belongs to the hydrocarbonate-calcium-magnesium hydrochemical facies. Below the outcrop of sulfate rocks, mineralization increases to 1179 mg/dm^3, and the composition of the water changes to sulfate-hydrocarbonate-calcium. The underground chemical runoff at the site is increased—1.12 g/(s km^2).

The average modulus of groundwater flow in the Irgina River basin is 10 l/(s km^2). Taking into account climatic indicators, this is the maximum possible specific value of underground runoff for the regions of the Middle Urals. The high modulus of groundwater flow is mainly due to the inflow of karst waters from the carbonate deposits of the eastern margin of the East European Platform. The inflow of groundwater from the Cis-Ural trough is observed below the Brekhovo village and is relatively small—several hundred l/s.

In the lower reaches of the Shurtana River, in a section about 5 km long, an intensive (940 l/s) inflow of groundwater was recorded.

Karst phenomena. The development of karst in the described region is associated with various types of groundwater: fissure-karst waters of carbonate deposits of the Ufimian swell, karst waters of sulfate deposits of the Kungurian stage, fractured-pore waters of terrigenous deposits of the Kungurian and Ufimian stages (Gorbunova 1965).

The karst of the region is of a different age. Probably, its development began in the Mesozoic. In the Paleogene-Neogene, large karst lacustrine depressions were filled with sediments. In the Quaternary period, in connection with the uplift of the Ufimian swell and the deepening of the Sylva River valley, karst processes revived. The southern part of the region experienced the greatest uplift.

In the Ust-Kishert village, Kovaleva (2010), based on survey materials, identified eleven types of geological sections (Fig. 3.102). In this area, sinkholes (205 forms) are most developed, including karst lakes of sinkhole genesis, and underground cavities (88 forms) (Fig. 3.103).

Most of the surface karst forms are confined to the areas of distribution of type III (27%) and type V (28%), while in the territories formed according to types VII–XI, practically no karst manifestations were found, which is explained by the absence of carbonate-sulfate Irene deposits. On the territory of type V, most of the lakes of karst genesis are located: Molebnoe, Yama, Bezymyannoe, Kisloe, Zapadnoe

Fig. 3.101 A pond in the village of Suksun, created in 1739, where hydrogen sulfide therapeutic mud is formed (*photo by* N. Maksimovich)

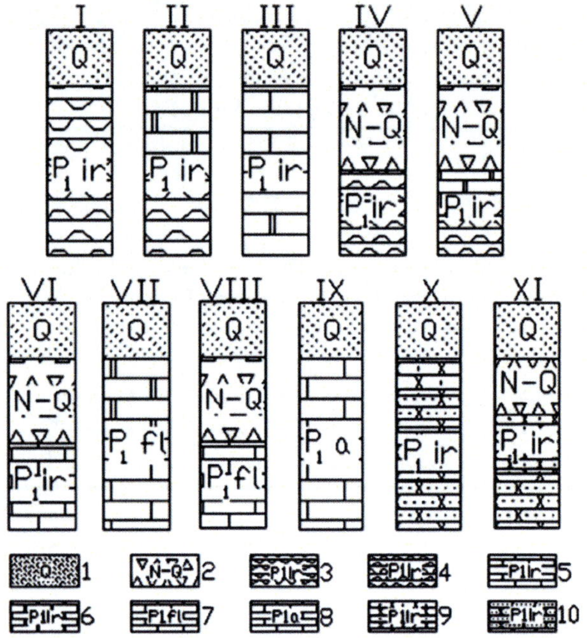

Fig. 3.102 Types of geological sections of the territory of the village of Ust-Khert (Kovaleva 2010) 1—Quaternary deposits; 2—Neogene-Quaternary deposits; 3—gypsum of the Irensky horizon; 4—anhydrites of the Irensky horizon; 5—limestones of the Irensky horizon; 6—dolomites of the Irensky horizon; 7—limestones and dolomites of the Philippovsky horizon; 8—limestones of the Artinsky stage; 9—siltstones of the Irensky horizon, 10—sandstones of the Irensky horizon

Berezinskoe, Srednee Berezinskoe, Vostochnoye Berezinskoe, and Vosmerka (Fig. 3.104). Underground karst cavities, as a rule, are confined to the sequence of alternating sulfate and carbonate Irenian rocks, overlain by Quaternary sediments (Kovaleva 2010).

North of the Sylva River, in the northwest of the line Podnikolskaya village—Posad-Kishert village, there is a strip of sinkholes, karst lakes, and kettles up to 2 km wide through the Shastino and Bezukladnikovo villages to the Karnaukhovo village on the Shakva River and further north. In the vicinity of the Ust-Kishert village, 160 sinkholes have been described. To the south, on the left bank of the Kishertka and Sedinka Rivers, on the slopes of the Dreminskaya and Burtsevskaya depressions, 150 sinkholes were recorded (Gorbunova 1959).

In the Kishertsky karst region, there are several areas corresponding to different stages of karst development (Gorbunova 1965). Karst and erosional-karst depressions are more mature forms of sulfate karst, they have a diameter of more than 1 km and are extended along the eastern part of the Ufimian swell.

The Nizkovskaya depression has a diameter of about 2 km. From the west and south, it is bounded by steep slopes, and, in the north, it opens into the valley of the Sylva River. The bottom of the depression is flat, and, in some places, it is swampy, in the east of the depression it slightly rises and passes into the valley of the Sedinka River. A large dry valley flows into it from the southwest.

Numerous ascending springs emerge at the bottom. The bottom of the karst depression is a kind of "window" where karst waters are unloaded. Their feeding area is the field of craters near the Dremino village and the Ufimian plateau. From here, the waters move north and east towards the depression, dissolving gypsum.

Most of the spring runoff (80%) comes from three springs. Of greatest interest are the two large ones on the right slope of the valley of the Kishertka River with a flow rate in low-water periods up to 20–25 l/s and a water temperature of 6.5 °C, which are characterized by an increased content of carbon dioxide. A sharp increase in groundwater

Fig. 3.103 Histogram of the frequency of occurrence of karst forms within territories with different types of geological section (Kovaleva 2010)

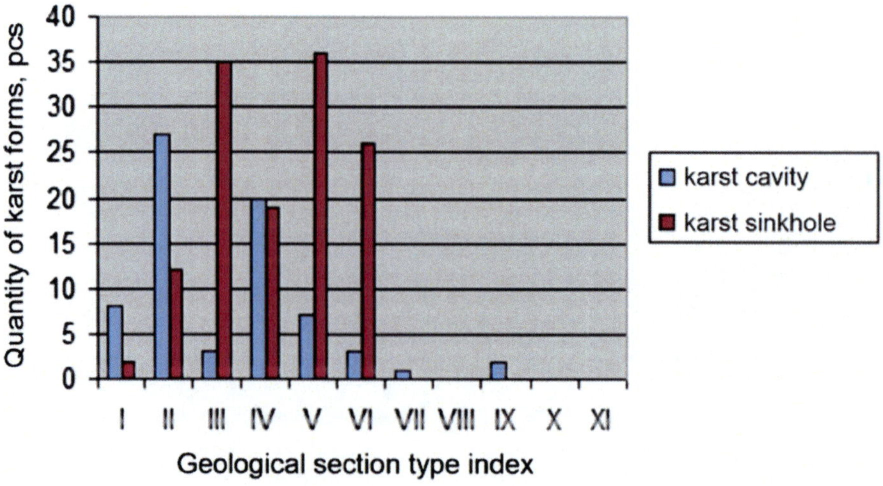

Fig. 3.104 Karst Lake Vosmyorka in the Ust-Kishert village (*photo by* N. Maksimovich)

flow in the extended linear zone of the Kishert dry valley, along with other hydrogeological indicators, indicates the existence of a local tectonic structure here (Fig. 3.99c).

The Kishertka River first lost its drainage role, apparently, in the middle section, below the Kulikovo village. The first inhabitants who founded the Kulikovo, Sabarka, and Morgunovo villages 150–200 years ago, in all likelihood, settled on the banks of the existing river, since there were no other surface water sources here. In the memory of the current generation, a surface watercourse to the Kulikovo village was observed. Isotopic studies have shown that the waters in the springs and the Kishertka River in the territory of the Nizkoye village are not related to the annual course of atmospheric precipitation. The delay of the isotope signal occurs with an interval of two months, which indicates the rate of passage of waters in the karst system from the place of feeding to the place of their discharge (Kazantseva et al. 2020).

The Dreminskaya and Burtsevskaya depressions are located to the south of the Nizkovskaya on the watershed. The Dreminskaya depression is located in the upper reaches of the ravine, which flows into the Sedinka River. The lowest part of it is occupied by a lake with fresh water. On the western and northern slopes, there are swallowing and silted funnels. In one of these plaster casts come out. In 1946, there was a failure in the Dremino village.

Atmospheric precipitation, absorbed by the funnels, is discharged as sources in the Nizkovskaya depression. The Burtsevskaya lacustrine depression is located between the headwaters of two ravines flowing into the Sedinka River. Its slopes are covered with numerous mostly dry sinkholes, and the bottom is a series of lakes and bogs with fresh water (Gorbunova 1957).

Mazuevskaya depression—one of the most interesting areas of gypsum karst development, located 2 km west of the Mazuevka village, is about 3 km long, 1.5 km wide, and up to 100 m deep. It belongs to the type of lacustrine karst depressions with lakes, kettles, ponors, springs, and karst rivers. The deepest parts of the bottom of the depression approach the level of karst waters. The flow rate of water in the Mazuevka River below the outlet of the springs reaches 500 l/s, and the salinity is 1.8 mg/dm^3. The bottom of the depression is swampy and occupied by lakes with mineralization from 0.07 to 1.6 mg/dm^3. According to the genetic classification (Maksimovich 1969), two types of lakes are distinguished in this territory: karst lakes in the zone of vertical descending circulation and lakes in the zone of horizontal circulation of karst waters. According to the

Fig. 3.105 Karst lake in the Sovetnaya village (*photo by* N. Maksimovich)

nature of feeding (Kataev 2000), all the lakes studied belong to funnel lakes of two types: with surface and mixed feeding (Krutik and Kilin 2018).

Sulfate karst waters play a significant role in their nutrition. In the Mazuevskaya depression, waters of the sulfate type (sulfate-sodium subtype), belonging to the gypsum and magnesium groups, prevail, which is explained by the distribution of rocks of the Koshelevsky and Popovsky formations, as well as deposits of the Irenian horizon. Waters with mineralization from 0.2 to 2.5 mg/dm^3 are widespread (Fetisov 2005). Covered and suballuvial types of karst predominate (Gorbunova et al. 1992). 2000 sinkholes, kettles, large lacustrine depressions, lakes, springs with a flow rate of up to 54 l/s, sulfate karst rivers, and small caves have been recorded here (Gorbunova 1965).

Karst breccia is encountered at the contact of the underlying Philippovsky formation. A strip of karst breccias up to 2 km wide and 40 km long stretches along the eastern wing of the Ufimian swell. They consist of angular fragments ranging in size from 1 to 50 cm or more. Fragments are mainly represented by dolomites, limestone, sometimes silicified, less often—anhydrite, and to the east of the Irgina River—sandstones and clays. The breccia thickness reaches 45 m (Lavrova 2005).

In areas of intensive karst development, the occurrence of rocks is disturbed due to the collapse of the arches of karst cavities. The angles of incidence can reach several tens of degrees. Karst dislocations are both folded and discontinuous (Maksimovich and Gorbunova 1958; Lukin 1964).

The bottom of the Suksun-Sovetskaya erosion-karst depression is occupied by a pond fed by the sulfate karst rivers Sovetskaya and Kiselevka and by the Klyuchi River, which originates from fresh sources of karst waters of the Ufimian swell at the mouth of the Suksun dry valley. The left-bank part of the Soyanka River valley and the south-eastern slope of the pond are heavily karsted. In the Sovetnaya village, there are several karst lakes with a diameter of up to 100 m (Fig. 3.105); most of them are at the stage of overgrowth.

The most common form of manifestation of karst processes on the territory of the Suksun settlement is karst sinkholes (about 430 forms). The density of karst forms here varies from several units to 100 or more forms/km^2 (Kovaleva and Nikolaev 2013).

Underground karst content is most studied in the Suksun village and its environs. The sulfate-carbonate deposits of the Irenian horizon are intensely karst. The depth of the roof of the Irenian karst sediments varies from 0.2 to 35.0 m. They are encountered by wells at absolute elevations of 160–190 m. Eleven karst cavities, filled with terrigenous sediments, were uncovered, among which crushed stone and marl gruss prevail, clay is met less often. The depth of the opening ranges from 11 to 45 m. The amplitude of most of the cavities is up to 2 m, the maximum is 12.5 m (Kovaleva 2011).

The development of karst is accompanied by the replacement of sulfate rocks by karst breccias, and this process spreads from the massif of carbonate rocks, involving new areas. The development of karst will apparently continue until the complete destruction of the sulfate strata. This process is facilitated by the ongoing uplift of the area (Gorbunova 1965).

The Dikoozerskaya depression is similar to the Burtsevskaya depression, but is at an earlier stage of development. Its bottom is occupied by fresh water lakes. Lake Dikoe (dimensions 400 × 140 m) has a depth of about 7 m (Fig. 3.106).

The lake in the Nizhnaya Odina village (dimensions 122 by 94 m; depth—19 m) is confined to the sinkhole that arose at the bottom of the lake basin in 1953. The sinkholes are still being recorded. In the spring of 2020, in the Dikoe Ozero village, two small sinkholes occurred in the area of the church. They are located in the path of the water flow absorbed by the ponor located 120 m to the southwest (Fig. 3.107). A deformed residential building is located nearby (Fig. 3.108).

Fig. 3.106 Karst lake in the Dikoe Ozero village (*photo by* N. Maksimovich)

Fig. 3.107 Karst ponor in the Dikoe Ozero village (*photo by* N. Maksimovich)

In 2014, a change in the chemical composition of the lakes was noted (Krutik and Ovsyannikova 2018) compared to the data of earlier studies. The change in facies is probably associated with the evolution of karst lacustrine basins, as a result of which cracks and pores that provide a connection between lacustrine and karst waters become clogged.

On the territory of the region, 49 dips with an average size of up to 7 m in diameter and up to 4 m in depth were recorded over 100 years, of which 5 dips with a maximum size of up to 50 m in diameter and up to 40 m in depth. The largest one in the Ust-Kishert village on the second terrace of the Sylva River occurred on August 28–29, 1949. First, a saucer-like depression appeared on the surface of the earth, in the northwestern wall of which a crack appeared. Then the collapse began, accompanied by noise and outpouring of water. Five hours later, the collapse resumed, which led to the formation of a second sinkhole; both filled with water. The two lakes ended up in a common basin and were separated by a ridge up to 1.9 m high. The length of the basin was 50.5 m, the width of most of it was 40 m, and the smaller one was 22 m (Gorbunova 1965). The deepest sinkhole occurred in the Brekhovo village in 1953; at the time of the survey, the sinkhole depth was 40 m.

Caves. According to the archival data of the Mining Institute of the Ural Branch of the Russian Academy of Sciences and the data provided by speleologists, there are 11

Fig. 3.108 Deformation of a house in the Dikoe Ozero village (*photo by* N. Maksimovich)

Table 3.5 Underground karst occurrences in the Kishertsky region (after Gerasimova 2020)

N	Name	Length, m	Amplitude, m
		According to the "Perechen peshcher Urala i Priuralya", 1992; "Peshchery Povolzhya", 2010 with additions	
1	Varsanofieva (Mazuevskaya, Burtsevskaya, Bely Kamen, Volchya Yama)	**205**	**23**
	Varsanofieva-2	66	
	Varsanofieva-3 (Lake)	50	
	Gorenka	200	
	Gribushinskaya		
	Ermakovskaya	10	
	Crystalnaya	15	
	Penkovskaya	5	
	Lobachevskaya		
2	Polyakovskaya	**43**	**4**
	Posadskaya	4	
Total number of underground karst occurrences—11		645	
Total number of caves, taking into account works of 2018–2019 years:—2		**248**	

underground karst occurrences in the Kishertsky region that are currently accessible to humans, in gypsum and anhydrite, less often in dolomites. The table lists them in alphabetical order; objects that are considered caves are numbered (Table 3.5). The rest of the karst forms are grottoes or inaccessible to humans.

Due to the high solubility of sulfate deposits and the influx of fresh water from the outside, the caves of this area are rapidly evolving.

Until the 1990s, two more caves existed within the Mazuevskaya karst depression: Varsanofieva-2 and Varsanofieva-3, but after the construction of the gas pipeline, the groundwater flow was disrupted and the caves were filled with clay deposits. In 1992, within the Mazuevskaya depression, the Kristalnaya and Gorenka caves were described (Lavrov 1992). At the time of research (August 2019), the entrance to these caves was blocked by avalanche sediments. A similar picture is observed for the Penkovskaya caves (the outskirts of the Penki village) and Yermakovskaya (1.2 km west of the Posad village).

Gribushinskaya, Lobachevskaya, and Posadskaya cavities are grottoes.

Thus, as of January 1, 2020, two caves are available for exploration within the Kishertsky region: Varsanofieva and Polyakovskaya (Gerasimova 2020).

The Varsanofieva cave (other names: Mazuevskaya, Burtsevskaya, Bely Kamen, Volchya Yama) is located in the Kishert district, within the Mazuevskaya karst depression, 4 km west of the Mazuevka village. It is located at the bottom of the Volchya Yama sinkhole, 120 by 90 m in size and 54 m deep. After a collapse in 2005, the entrance was heavily heaped up with boulders at the base of a 20 m high rock outcrop in the northeastern part of the sinkhole. It was first surveyed by V. A. Varsanofieva in 1911. A detailed map and description of the cave were compiled by Permian cavers in 1992 (Fig. 3.109).

The cave was formed in the gypsum anhydrites of the Popovsky formation of the Kungurian stage and is a system of passages with small grottoes at different levels.

In the lower part of the grottoes, there are deep cavities with water. The length of all passages reaches 205 m, and the depth of the cave is 23 m (Gerasimova et al. 2020).

Ice deposits in the cave were described by Varsanofieva (1915), Gorbunova (1965), and Lavrov (1992). In 2019, the cave also contains perennial ice and sublimation crystals.

The Polyakovskaya Cave is located 2.3 km northwest of the outskirts of the Posad village, on the left side of the Posad Log of the right bank of the Sylva River.

According to the semi-instrumental topographic survey in 2019, the total length was 43 m, and the amplitude was 6 m (Fig. 3.110; Gerasimova et al. 2020).

In the immediate vicinity of the rock mass of gypsum anhydrites, which includes the cave, there is a ponor lake, which is filled with water only in the spring.

Previously, the entrance was located at the base of the northeastern part of the rock massif, and the cave was described as having a length of 200 m (Caves of the Volga … 2010). At the time of the research (November 2017), the described entrance was blocked by blocky-gravelly talus, and 50 m to the south, also at the base of the gypsum anhydrite outcrop, a new entrance, 1.3 m wide and 1.0 m high, appeared.

The cave is a system of passages developed along the cracks of the sublatitudinal and sub-meridional directions, at the intersection of which, as a result of the action of landslide processes, small grottoes are formed.

Existing and proposed specially protected natural areas

Existing SPNAs. Within the Kishert district, there are two protected areas associated with sulfate karst of regional and local significance, with a total area of 112.3 ha: Bely Kamen and Lake Karasye. They were offered protection by K. A. Gorbunova in 1959.

The Bely Kamen is a landscape natural monument of regional significance with an area of 90.3 ha with a protected zone of 55.1 ha and is located 1.5 km southwest of the Krokhalevo village. The territory is a classic karst depression with five lakes: Maloe, Svetloe, Chornaya Yama, Svetlaya Yama, and a lake with a drift (Specially Protected … 2017). The depression is characterized by the predominance of underground runoff of sulfate–calcium waters. The Kristalnaya and Gorenka caves were previously described within the SPNA in gypsum outcrops in the marginal parts of the depression (Lavrov 1992), but they are not available today.

Lake Karasye is a geological natural monument of regional importance with an area of 22 ha and is located 500 m southwest of the Krokhalevo village (Fig. 3.111). The lake with drift, 100 by 500 m in size and 5.5 m deep, like the SPNA Bely Kamen, is located within the Mazuevskaya depression.

Proposed SPNAs. Volchya Yama sinkhole with Varsanofieva cave, Zuevsky spring, Lake Proval (Kishert district), Suksunsky pond, and Lakes Krugloye, Nizhneodinskoe, and Molebnoe in Suksun district, according to the authors, were unreasonably excluded from the SPNAs after 2009. It is necessary to restore the protected status of these objects, in addition, to submit for consideration a unique section of the groundwater discharge zone of the Kishertka River at the junction of carbonate and sulfate rocks in the Nizkoye village. In total, about 18 springs come out in the area of the village, the average flow rate of which is 4 l/s.

Economic activity and environmental pressure. Currently, on the territory of the Kishertsky karst region, there are mainly enterprises that are engaged in crop production and processing of agricultural products, there are railways, pipelines, mining of minerals, mineral waters, etc.

The Sverdlovsk railway has a technogenic impact on the territory of the region. According to the assessment of the possibility of karst deformations of the base of the railway bed (Maklashin and Alvanyan 2015), the interval from 1535/36 km (Kungur station) to 1556/57 km (Ust-Kishert station) is subdivided into segments with the corresponding

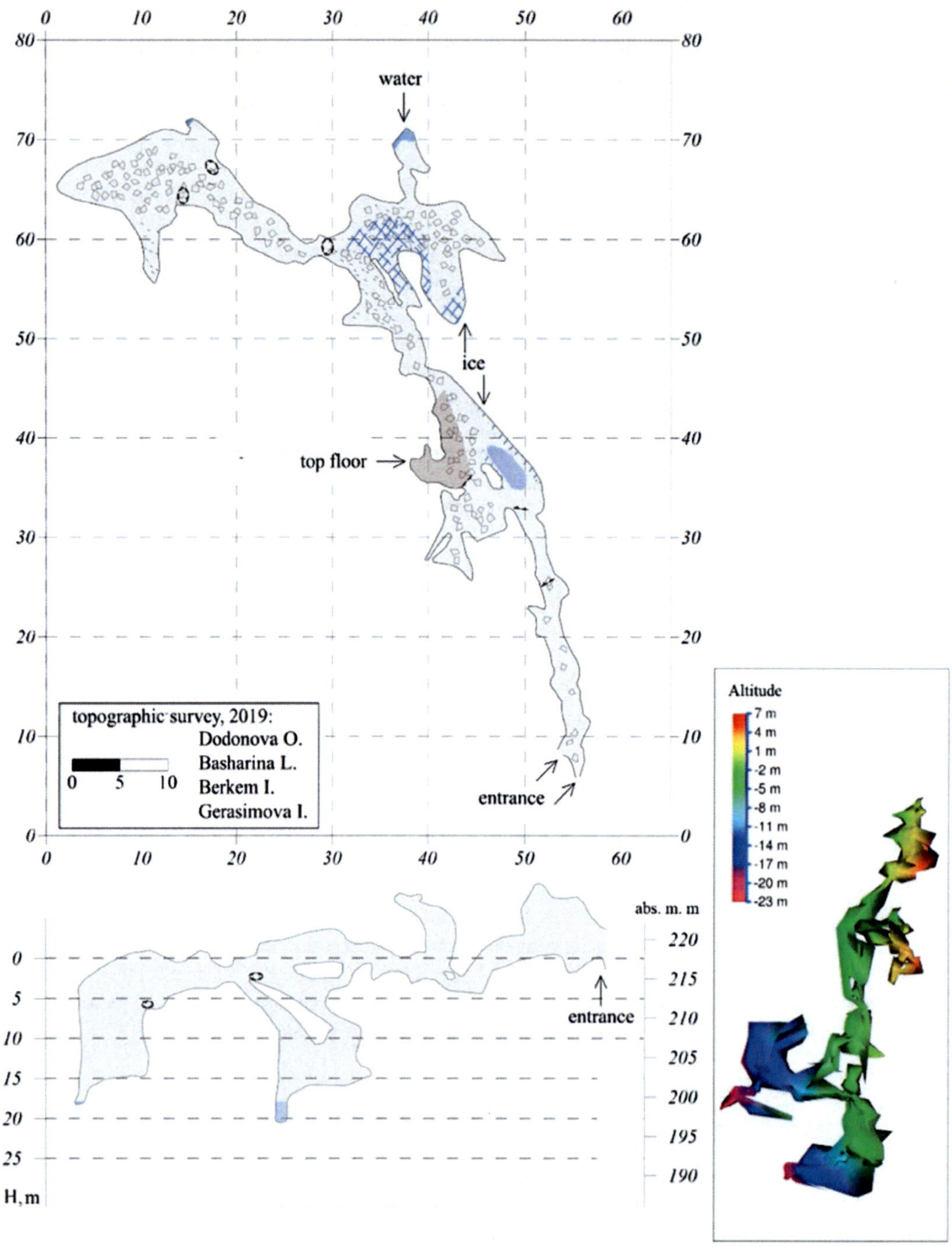

Fig. 3.109 Plan and section of the Varsanofieva cave (Dodonova, Basharina, Berkem, Gerasimova 2019)

category of stability, determined by specific set factors of karst hazard. Based on the results of the analysis of the data of the geological and geophysical complex of works, 15 potentially dangerous in karstological terms were identified, the degree of danger of which varies from II to IV (medium, high, and very high). The total length of karst-hazardous areas at the base of the bed is 7223 m, of which with a medium and high degree—4578 m and with a very high degree—480 m according to karstological data, and 2640 m—according to geophysical and engineering-geophysical data.

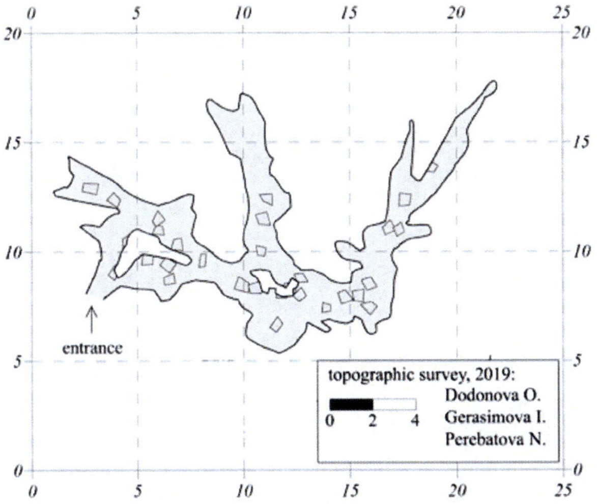

Fig. 3.110 The Polyakovskaya cave plan (Dodonova, Gerasimova, Perebatova 2019)

In 2013, the authors found subsidence of the soil near the Zuevsky spring near the railway in the Ust-Kishert village (Fig. 3.112).

In the Bolshoi Log open sinkhole within the Kishert dry valley, construction crushed stone, limestone, and dolomite are mined (Fig. 3.113).

Near the Mazuevskaya karst depression, there is the largest **strontium ore deposit** in Russia with an area of about 20 km^2, with approved reserves of the upper ore-bearing horizon of 15.2 million tons of ore; currently not being developed (Geological Monuments ... 2009).

The Shvalevsky Log **deposit of stucco** is located in the Kishert district near the Posad village. In the area of the deposit, karst processes are developed, expressed on the surface in the form of logs and numerous karst sinkholes. The Bobinskoye gypsum deposit is also located in the area.

Deposits of medicinal mud. The surface waters of the areas of development of sulfate karst are enriched in sulfates.

Fig. 3.111 Karst Lake Karasie with drift on the surface (*photo by* O. Kadebskaya)

Fig. 3.112 Subsidence of soil near the Ust-Kishert village (*photo by* N. Maksimovich)

In conditions of oxygen deficiency, sulfates with the participation of microorganisms are reduced to hydrogen sulfide, which leads to the formation of therapeutic mud.

The Suksunsky pond was created in 1739 on the Suksunchik River simultaneously with the founding of the Suksunsky copper smelter. In the shallow bay in the southern part of the pond, gray and black mud with a pungent smell of hydrogen sulfide with small plant impurities is widespread. They belong to freshwater medium-ash algal-putrefactive sapropels with good medicinal properties (Table 3.6). The specific gravity of mud is 1.28 (Maksimovich and Chistyakov 1957). The dirt smells like hydrogen sulfide. The color is black, the contamination is significant, the consistency is soft, and the moisture content is 47.8%. The mud of the Suksunsky pond, according to the classification of Yarotsky (1956), belongs to the lake-spring type.

The mud of the Suksunsky pond is characterized by high saturation of hydrogen sulfide and dispersion, has a high colloidal complex, in which the crystalline skeleton prevails, with an increased content of silicate. Organic substances are represented by bitumen, humins, cellulose, compounds of nitrogen, phosphorus, iron, sulfur, the remains of algae, and microorganisms. The mud also contains biologically active substances, enzymes, hormone-like compounds, and trace elements. The composition of the Suksun mud is similar to the mud of the Soltsy resort, silt mud, and silt-peat mud of the Lipetsk resort (Maksimovich and Chistyakov 1957). The Klyuchi resort has been provided with mineral balneological resources, including mud, for exploitation for more than 100 years (Belov et al. 2004).

This area is also characterized by accumulations of silt hydrogen sulfide mud in karst lakes and in wide swampy river valleys, for example, in the Malaya Irgina Kiver.

Mineral waters of sulfate–calcium composition are widespread in the area of development of the region. The curative-table waters of sulfate–calcium composition are confined to karst gypsum-anhydrite deposits of the Kungurian stage. They open up at depths from 40 to 100 m.

The Klyuchi resort is located here, where two types of mineralized waters and curative silt mud are used for treatment, there are geological prerequisites for the discovery and introduction into the practice of medical procedures of low-mineralized waters with an increased content of water-dissolved organic matter, iodine-bromine brines with an increased content of iodine, bromine, boron, and other microcomponents. Well 1/92 (34.5–117.5 m) is used for medical purposes at the Klyuchi resort. Klyuchi water, which is an analogue of the Smirnovskaya mineral water,

Fig. 3.113 Bolshoy Log Quarry in Kishert dry valley (*photo by* O. Kadebskaya)

Table 3.6 The chemical composition of the aqueous and hydrochloric acid extracts of mud from the Suksunsky pond (Geological monuments … 2009)

Water extract, mg/dm^3	H_2S	CO_2	Ca	Mg	Na	HCO_3	SO_4	Cl
	137.2	4456.0	372.7	64.7	225.6	264.7	663.3	553.0
Hydrochloric acid extract, g/100 g	SiO_2	$Fe_2O_3 + Al_2O_3$	Ca	Mg	PO_4	P_2O_5	SO_4	Cl
	0.0409	3.067	3.678	0.4678	0.0162	0.0121	0.2356	202

has a salinity of 2.45 g/dm^3 and a pH of 7.3 and has been used for bottling since 1995 (Geological monuments … 2009; Egorov et al. 2007).

Outlets of hydrogen sulfide waters are located near the Bolshiye Klyuchi village, Suksun district. The mineral composition is different: sulfurous ferruginous and saline. The springs come out on the earth's surface at a distance of 1–6 m from each other and are located mainly on the right bank of the Irgina River branch. Two hydrogen sulfide springs have survived: one—at well 1/52, the second—at Gorodishche Mountain. The water of the springs belongs to the group of low-mineralized waters of the sulfate–calcium hydrochemical facies with an increased content of sodium and chlorine (Geological monuments … 2009). The industrial bottling of mineral medicinal table waters Klyuchi is carried out.

3.8 Osintsevsky Region of Covered Salt and Sulfate Karst

The region, of an area of 785 km^2, is confined to the arch of the Tulumbassko-Tisov swell, which is distinguished by an elevated position of the base of the Kungurian stage (Gorbunova et al. 1992; Shumilova and Maksimovich 2008). Administratively, it is located on the territory of the Berezovka, Kishert, Suksun districts of the Perm Region (Figs. 3.114 and 3.115).

In zones of increased fracturing, lenses of gypsum and salts karst, occurring among non-karst sediments. Here, in a strip from the Asovo village to the south through the Plotnikovo, Osintsevo, Ivankovo, and Tis villages, karst sinkholes (Fig. 3.116), subsidence, kettles, karst lakes, as well as underground forms of karst, are traced.

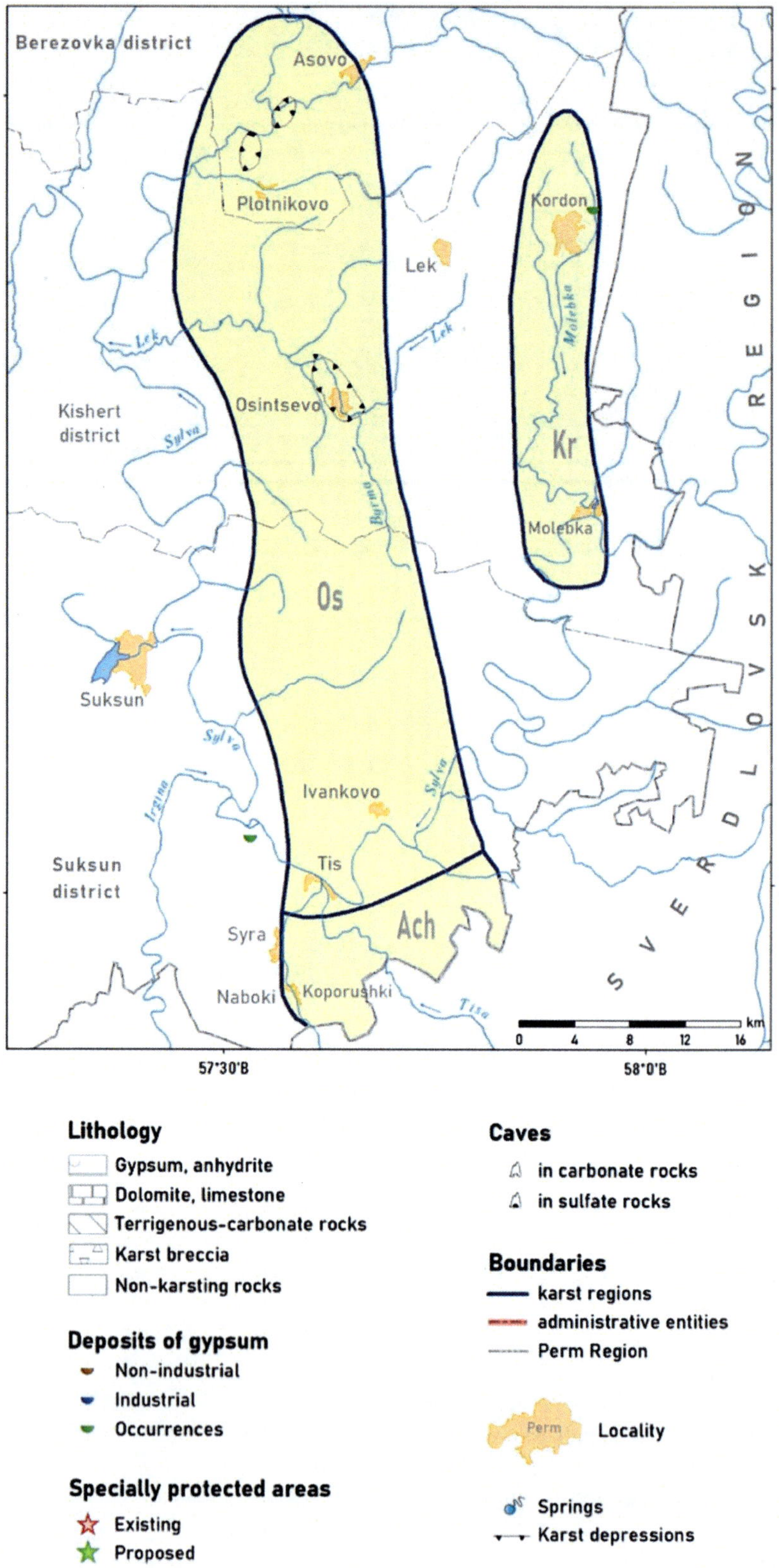

Fig. 3.114 Osintsevsky (Os), Achitsky (Ach), Kordonsky (Kr) karst regions

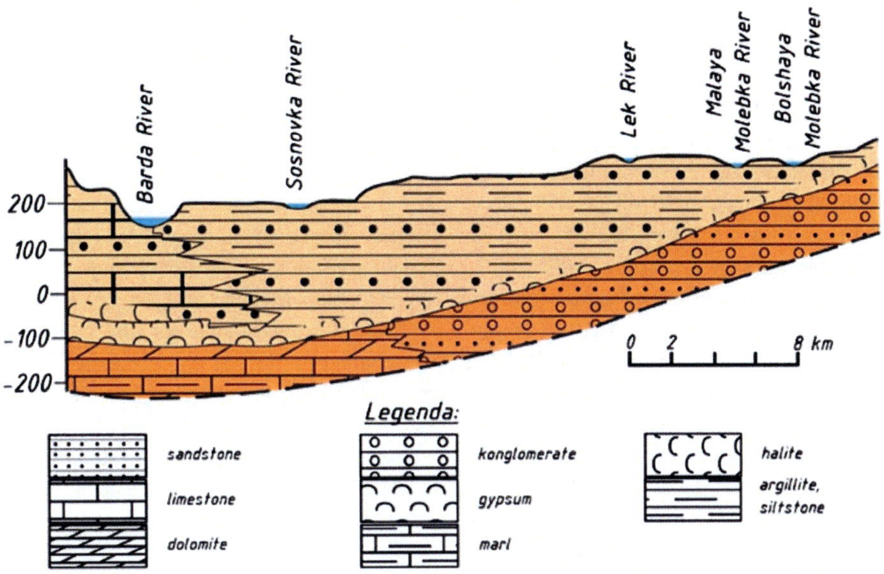

Fig. 3.115 Geological conditions for the development of karst in the Osintsevsky, Achitsky, and Kordonsky regions (Gorbunova et al. 1992)

Fig. 3.116 A sinkhole northeast of the Osintsevo village (*photo by* N. Maksimovich)

The funnels are filled with water in places, which leads to the formation of lakes. In river valleys, brackish springs flow out. The Plotnikovsko-Sosnovskaya and Osintsevskaya boggy depressions are 5–7 km in diameter.

According to the data of structural exploration drilling in Tulumbasovka area, zones of complete absorption of drilling fluid were noted in the area of the Asovo village at depths of 324–444 and 309–325 m (Shimanovsky 1973).

In the Byrma River basin, in the areas of sulfate karst, the fractured ground waters of the Koshelevsky sandstones have an average mineralization of 656 mg/dm^3, and the average content of sulfate ions is 275 mg/dm^3, while, in the Molebka River basin, where karst phenomena are absent, these values are 227 and 7.4 mg/dm^3, respectively (Ikonnikov 1975).

The discharge of saline chloride waters activates gypsum karst (see Chap. 1).

On the territory of the region, 12 km to the south of the Tulumbasy railway station (near the Osintsevo village), there is the Osintsevskoe drywall deposit. It is located in the valley of the Lek River. The useful stratum is a sheet-like deposit, folded by alternating layers and lenses of drywall and peat, with an average thickness of 3 m. The balance reserves of the category are 8.5 million tons. A peat-limestone stratum 0.5–7 m thick (average—3 m) is covered by a soil-vegetation layer and loams 0.2–2.2 m thick (average—0.7 m) (Vodolazskaya et al. 2015).

3.9 Achitsky Region of Sulfate Karst and Klastokarst

South of the Osintsevsky karst region, there is the Achitsky region of covered sulfate karst and klastokarst, mainly in the Sverdlovsk region. The area of the region on the territory of the Perm Region is 96 km^2. Administratively, it is located on the territory of the Suksun district of the Perm Region (Figs. 3.114 and 3.115; Gorbunova et al. 1992; Shumilova and Maksimovich 2008).

The karst of the region is due to the activity of the Kungurian terrigenous aquifer.

The slopes of the Syra River valley, along the banks of which conglomerates with carbonate-gypsum cement are exposed, are covered with choked dry funnels (Fig. 3.117; Gorbunova et al. 1992).

On the western outskirts and in the central part of the Syra village, there are numerous karst lakes, the largest of which: Lubenoye, Simenchi and Burlivoe (Fig. 3.118)—are in different stages of eutrophication.

Chains of small karst sinkholes are observed near the lakes. The largest craters are located in a field near the northwestern outskirts of the Naboka village: round—30 m in diameter (Fig. 3.119) and oval—30 by 70 m in diameter.

In the area on the eastern outskirts of the Bykovo village (Sverdlovsk region), a chain of 15 karst sinkholes is confined to the layer of gypsum and anhydrite (Chuvashov et al. 1990).

Areas of sulfate groundwater with a salinity of 1–3 g/dm^3 are confined to the lenses of conglomerates and conglomerate-breccias. Such sites are noted near the Koporushki village in the valley of the Syra River, and to the southeast—in the Sverdlovsk region in the valley of the Zyurya River near the Podgornaya village.

The gypsum content of the Sheshminsky sandstones determined the presence of brackish sulfate waters in their poorly washed part below the erosional incision. The springs flowing out in the southeastern outskirts of the Koporushki village in the valley of the Syra River have a flow rate of 17 and 60 l/s, and the spring in the Bykovo village (Sverdlovsk region)—about 180 l/s (Shimanovsky and Kuntz 1986). The spring on the right bank of the Syra River in the village of the same name is blocked off by a dam and is used as a fire reservoir (Fig. 3.120).

3.9.1 Kordonsky Region of Predominantly Covered Sulfate Karst

The region of an area of 172 km^2 is located in the Molebka River basin near the Kordon and the Molebka villages. Administratively, it is located on the territory of the Kishert district of the Perm Region (Figs. 3.114 and 3.121; Gorbunova et al. 1992; Shumilova and Maksimovich 2008).

The rocks are monoclinal (dip angle 1°–3°). Anhydrites with a thickness of 20 m were penetrated by a well near the Lek village at a depth of 130 m under sandy-clayey rocks (Fig. 3.115). In the vicinity of the Kordon village, they occur near the surface. In the center of the northern part of the village, there is a karst ravine with numerous sinkholes (Fig. 3.122), which are also observed throughout the entire village (Fig. 3.123), some of them are filled with water.

To the west of the village, karst sinkholes, mostly cone-shaped with a diameter of up to 50 m and a depth of 15 m, form a continuous strip. In the largest of them, gypsum is exposed.

According to local residents, in 1900, 1.5 km south of the Kordon village, an oval sinkhole appeared, 13 m wide, 22 m long, and about 14 m deep (Gorbunova 1954). The walls of the sinkhole are steep; they contain layered sandstones that lie almost horizontally. The sandy-marl stratum of the Kungurian stage with subordinate deposits of gypsum and anhydrite is developed here. In 2009, near the railway station in the garden of a private house No. 21 on the street Pervomayskaya, a karst sinkhole with a diameter of 10 m occurred, at a depth of 8 m, there was water in the sinkhole.

Fig. 3.117 Karst landscape of the Achitsky region (*photo by* N. Maksimovich)

Fig. 3.118 Karst Lake Burlivoe on the western outskirts of the Syra village (*photo by* N. Maksimovich)

Fig. 3.119 A sinkhole northwest of the Naboki village (*photo by* N. Maksimovich)

Since the hole was formed in the immediate vicinity of the railroad bed, it was filled up; thereafter, no visible subsidence was observed at the place of the sinkhole.

Karst funnels and sinkholes in this region are formed due to the leaching of readily soluble rocks underlying the sandstones.

3.9.2 Irensky Region of Predominantly Sulfate and Carbonate-Sulfate Karst

Irensky karst region is a classic example of sulfate karst, occupies the right-bank part of the Iren River basin, and its area is 3815 km^2. From the east, it is limited by carbonate rocks of the Philippovsky horizon of the Kungurian stage. The western border is drawn along the left bank of the Iren River, where the Irenian horizon plunges to the west and is overlapped by the Solikamsky horizon and then the terrigenous Sheshminsky one. In the north, the latitudinal section of the Iren River is conventionally taken beyond the border with the Nizhnesylvinsky karst region (Fig. 3.124). Administratively, it is located on the territory of the Chernushka, Kungur, Oktyabrsk, Orda, Suksun, and Uinsk districts.

Geological conditions for the development of karst. The region is located on the eastern edge of the East European Platform. It is located on the border of two large structures: the Bashkir arch and the Bym-Kungur depression, complicated by small and medium-sized structures. One of them is the Ufimian swell (macrobrachyanticline). The Irensky karst region occupies the western wing of the swell, which is flatter than the eastern one (Fig. 3.125; Gorbunova et al. 1992). Linear erosional-karst relief elements are controlled by faults of the IV–V orders (Fetisov 2005).

The zone of active water exchange and karst formation is composed of rocks of the Kungurian stage. At the base lies the Philippovsky horizon of platy dolomites, often calcareous, oolitic, interlayers of chalk-like, in places cavernous. The caverns are filled with gypsum and anhydrite (Gorbunova et al. 1992).

Hydrogeological conditions. The Iren River is a left tributary of the Sylva River. The river is 214 km long, the basin area is 6110 km^2. The river belongs to the Eastern European type of water regime with a pronounced spring flood and mainly snow supply. The average long-term water discharge 15 km from the mouth (Shubino village) is 40.7 m^3/s (flow volume is 1.285 km^3/year). On average, about 45% of the annual runoff passes during the flood, the

Fig. 3.120 Karst spring in the Syra village, used as a fire reservoir (*photo by* N. Maksimovich)

Fig. 3.121 The Molebka river near the village of Molebka (*photo by* N. Maksimovich)

Fig. 3.122 Karst ravine in the Kordon village (*photo by* N. Maksimovich)

Fig. 3.123 A sinkhole in the northern part of the Kordon village (*photo by* N. Maksimovich)

average annual flow rate of the spring flood is 346 m^3/s, with a maximum of 596 m^3/s. The average discharge in the summer-autumn period is 21.8 m^3/s, and in the winter low-water period—15.2 m^3/s.

The river flows practically along the border of karst and non-karst rocks, numerous left tributaries (Turka, Syp, Bym, etc.) are fed by sources of sandy-clay deposits, right ones (Teles, Kungur, Tyush, etc.)—by karst springs, some of them several times they go into karst sinkholes and then reappear on the surface. The river flowing in the karst region in many areas is characterized by the presence of an under-channel flow and water outflows from it, which are observed in winter in the form of non-freezing areas (Fig. 3.126).

The zone of active water exchange is associated with ground waters of eluvial and alluvial deposits, waters of sporadic distribution of the Solikamsky horizon of the Olkhov breccia, and the aquifer of the Irensky horizon of the Kungurian stage.

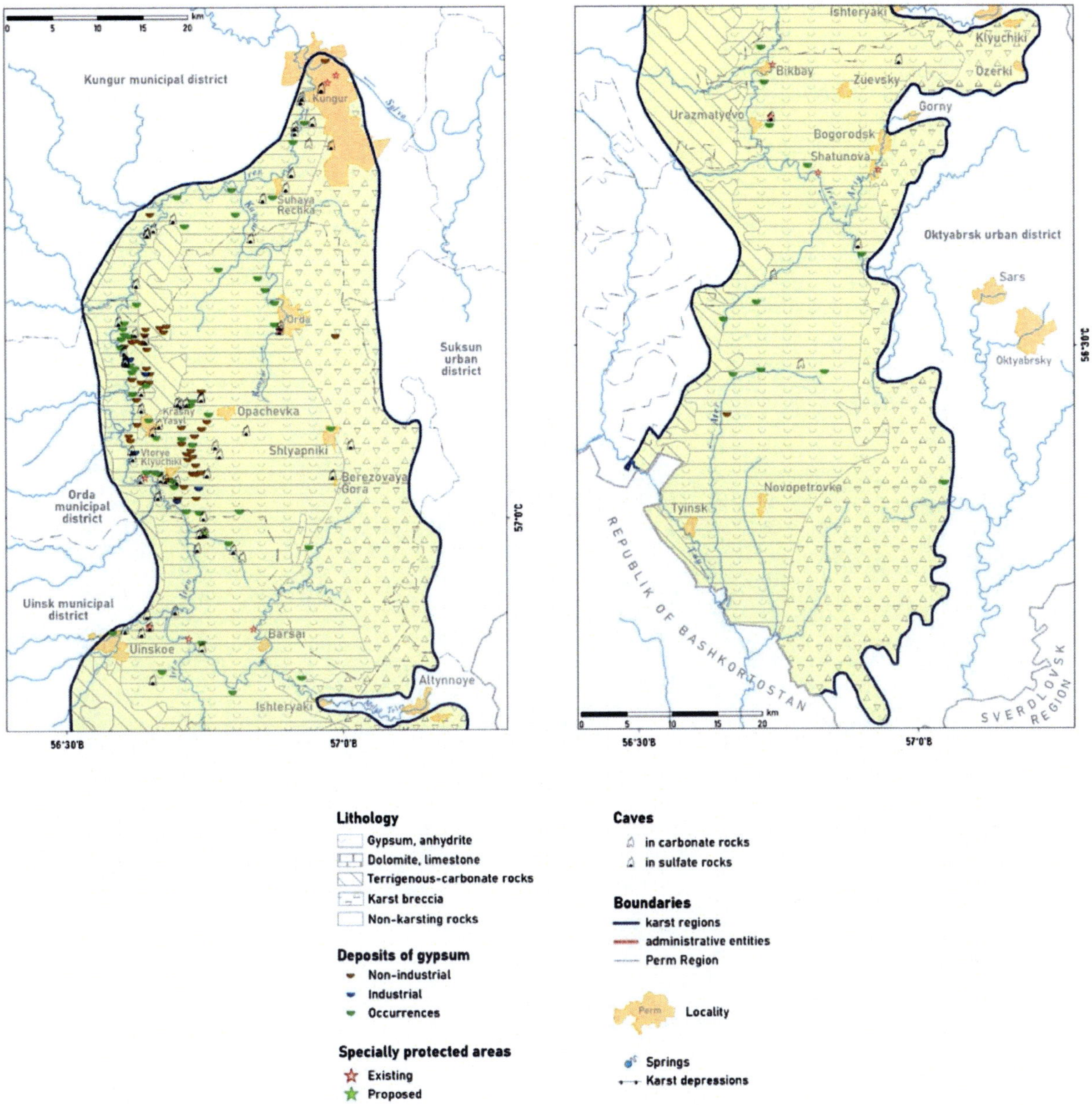

Fig. 3.124 Irensky karst region

In the Solikamsky horizon, the uneven distribution of waters is caused by the fracturing of the Irensky gypsum and anhydrite. In the disturbed areas, a latent overflow of water appears from the Solikamsky horizon to the Irensky horizon.

The Irensky aquifer is confined to highly karst gypsum-anhydrite and carbonate layers. In the Irensky horizon, both isolated karst watercourses and aquifers are developed (Gorbunova et al. 1992). Karst rivers are typical for the Irensky karst region (Churakovka, Pavlovka, Yasylka, Melenka), most of which are right tributaries of the Iren River.

Water-abundant zones (Mikhailov and Oborin 2006) with the discharge of fissure-karst waters of more than 10 l/s—Ordinskaya and Verkhnekungurskaya—are located in the Orda village and its environs. The main drain of both sections is the Kungur River. The largest karst springs with a flow rate of more than 300 l/s—Arsenovsky (Fig. 3.127) and Podzuevsky (Fig. 3.128).

The Ishteryak water-abundant zone is located mainly in the basin of the Maly Teles River and partly in the basin of the Aria River. The length of the section is 15 km, the width is up to 7 km, and the area is 92 km^2. The Ishteryak area is

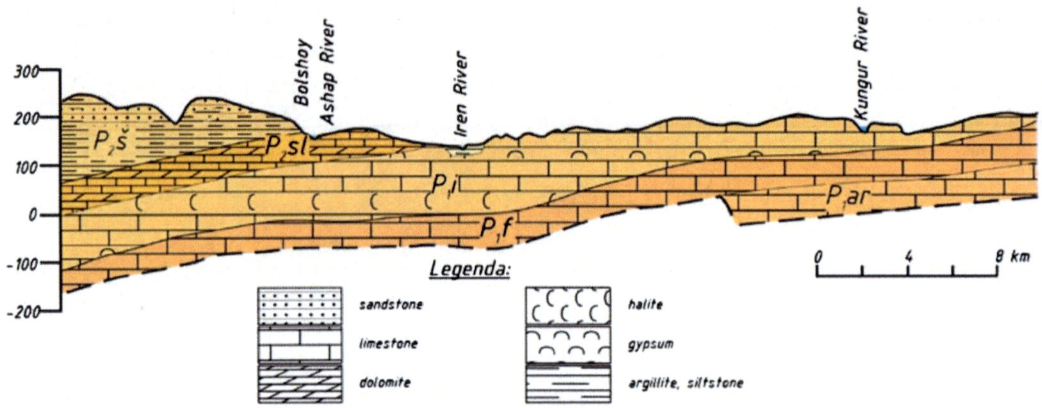

Fig. 3.125 Geological conditions for the development of karst in the Irensky region (Gorbunova et al. 1992)

Fig. 3.126 Outcrops of sulfate rocks in the area of the Ust-Turka village (*photo by* O. Kadebskaya)

confined to the intersection of the Kungur longitudinal zone of intense fracture faults with the emerging transverse zone of neotectocic deformations. A characteristic feature of the site is the presence of the Olkhovskaya karst breccia in the zone of descending karst circulation. The zone of horizontal circulation of karst waters is composed of rocks of the Philippovsky horizon.

Due to the intense and uneven tectonic fracturing of rocks, the flow of groundwater in the Ishteryak zone is sharply localized. The most powerful groundwater outflows are observed in the valley of the Maly Teles River in the Ishteryaki village and on the western outskirts of the Gorny village. On the outskirts of the Barsai village, there is a hydrogen sulfide spring, which is considered curative by local residents.

Fig. 3.127 The Arsenovsky karst spring in the Orda village (*photo by* N. Maksimovich)

Fig. 3.128 The Podzuevsky karst spring near the Podzuevo village (*photo by* N. Maksimovich)

In the valley of the Maly Teles River, on a 2.8 km long section, the inflow of groundwater is 850 l/s. The average specific water inflow into the drain in the Ishteryaki village was 300 l/(s km^2). The largest springs are located on the eastern outskirts of the village, where the Maly Teles River, having passed 6 km underground, again comes to the surface. The flow rate of the Maly Teles River at the outlet is 205 l/s. In places where sub-valley karst streams are unloaded, there are groups of springs with a total flow rate of up to 120 l/s.

The groundwater flow module in the Ishteryak zone in the Maly Teles river basin is 10–11 l/(s km^2). On the southern outskirts of the site drained by the Aria River valley, the groundwater flow module is 5 l/(s km^2). Here, on the western outskirts of the Gorny village, there is a large, concentrated outlet of groundwater with a flow rate of 160 l/s, associated mainly with tectonic fractures of the southwestern strike.

The underground chemical runoff in the central part of the Ishteryak zone is increased. On the one hand, this is explained by the structural-hydrogeological setting, on the other, by the presence of soluble sulfate rocks contained in the Olkhovskaya karst breccia.

The indicator of underground chemical runoff in the characterized part of the basin of the Maliy Teles river is 5.0–5.5 g/(s km^2). The discharge of sulfate ions here is equal to 1.5–1.6 g/(s km^2). On the southern outskirts of the site, where hydrocarbonate-calcium-magnesium waters with a salinity of 0.47 g/dm^3 are located and the role of sulfates in the composition of groundwater is small, the index of underground chemical runoff decreases to 1.65 g/(s km^2).

According to hydrogeological and hydrochemical data, a relatively large tectonic uplift of the brachyanticlinal type is clearly distinguished in the Ishteryak zone, which is actively developing at the present stage of geological history (Mikhailov and Oborin 2006).

The Pavlovsk water-abundant zone is confined to the Pavlovka River, the right tributary of the Iren River, 6 km long, formed from the confluence of two karst springs. Until the 20 s of the twentieth century, it was called the Turaevka River. The renaming of the village in 1922 entailed a change in the name of the river. For 1 km, it flows over the surface and then is lost under an outcrop of gypsum on the right slope of the valley. The underground section is 1.2 km long and ends with a powerful karst spring. The underground course of the river can be traced in its middle course in two funnels.

Mineralization of water is from 0.1 to 3.1 g/dm^3. They belong to the sodium sulfate subtype (Fetisov 2005). Sporadically watered karst breccias are common in the eastern part of the area at the border with the Philippovsky carbonates.

Karst phenomena. Karst on the territory of the region is characterized by high intensity, high densities of karst forms, significant coefficients of surface areal and deep karst content, and a high frequency of sinkholes. Karst refers to bare in areas of rock outcropping, choked, subeluvial, suballuvial, in places covered—under the Solikamsky non-karsting or less karsting carbonate deposits (Gorbunova et al. 1992).

The forms of manifestation of karst are varied. Karren is developed on the exposed gypsum slopes of the valleys of the Iren, the Aspa, and the Sudinka Rivers. Ponors in the vicinity of the Krasny Yasyl village are gaping holes in funnels. In the place where the Sudinka River disappeared, they look like cracks. Karst wells and mines in gypsum of the Lunezhskaya suite are known in the interfluve of the Aspa and Iren Rivers. The predominant forms are sinkholes that form karst fields. The highest density of sinkholes is observed on the slopes, on the slopes of valleys, and ravines, reaching values of 263 forms/km^2 (Pavlovskoe karst field). On an area of 19 km^2 of the Krasnoyasyl field, 2537 sinkholes were recorded (Fetisov 2005). Coefficient of areal karst content for individual fields is 21%. On the area where the Demidkovskaya suite of gypsum develops in the vicinity of the Arsenovka and Zakharovka villages at the watershed and on the slope of the Opachevka River valley, the density of karst forms reaches 160 forms/km^2. Funnels are confined to the slopes of the valleys and the slopes of watersheds of the Kungur, the Opachevka, the Sudinka, the Bolshoi, and the Maly Teles rivers. In the direction from the valleys to the watersheds, their density decreases, which is explained by an increase in the thickness of the cover deposits in the same direction. In places where gypsum and anhydrite occur shallowly, or come to the surface, the density of the sinkholes increases sharply. Large sinkholes reach 100 m in diameter and have a depth of up to 30 m. Chains of sinkholes in a number of places are confined to fractured zones. In the valley of the Sudinka River, they are located parallel to the river in accordance with the strike of the cracks of the onboard resistance. In the Teles river valley, the fields of linear funnels develop along tectonic faults (Gorbunova et al. 1992).

For the zones of gypsum distribution, erosional-karst and karst ravines are characteristic, often with outliers along the sides (Fig. 3.129).

The blind ravine south of the Verkhniy Kungur village, about 400 m long and 50 m wide, is perpendicular to the Kungur River valley, but the flow in it does not go to the valley, but in the opposite direction—to the funnel. There are similar logs near the Krasny Yasyl, Orda, and Pavlovo villages. Karst basins are common in gypsum and karst breccias. Some of them, the kettles to the west of the Medyanka and Suda villages, are stretched along the exit of the plaster packs, others, near the Shlyapniki village, are

Fig. 3.129 Gypsum remnants west of the Arsenovka village (*photo by* N. Maksimovich)

rounded in plan. Karst depressions are formed, as a rule, at the boundary of the Philippovsky carbonates and gypsum-anhydrite deposits of the Irensky horizons. These are vast depressions with flatter eastern and steep western slopes, coinciding with the boundary of gypsum distribution, and are distinguished by a large thickness of karst breccias. Such are the depressions southeast of the Sukhaya Rechka village, south of the Arsenovka village, near the Shlyapniki village, the Ozerki village, and the Medyanka village. Depressions contribute to the concentration and infiltration of surface runoff, feeding of karst waters that flow out on contact with gypsum, giving rise to the Medyanka, the Ariy, and others rivers.

In the Irensky karst region, there are more than 500 karst lakes, the largest of which is Lake Shchuchye in the Oktyabrsk district. In the spring of 2019, the water in the lake almost completely went into a karst sinkhole that arose in the central part of the funnel (Fig. 3.130). In 1950, a lake disappeared in the village of Shlyapniki (Orda district). In the Pochinki village, the water from the lake went underground several times.

Karst processes can be accompanied by sinkholes (Fig. 3.131) and subsidence, which pose a serious danger to civil and industrial facilities. Lukin and Ezhov (1980), as a result of engineering and geological studies in the territory of the Krasny Yasyl village, outlined five categories of stability of karst areas. The watershed of the Yaslash and the Iren Rivers, where the strata of the Solikamsky gypsum rocks have been preserved, is classified as a stable area. Water infiltrated through it becomes less aggressive in relation to gypsum and anhydrite, which excludes the formation of large cavities. Possibly slow subsidence of Solikamsky rocks over places of gypsum leaching. The relatively stable area, confined to the bottom of the Yasyl River valley, is characterized by the occurrence of gypsum and anhydrite under sandy-clay deposits below the level of mineralized groundwater. It is possible that subsidences up to 5 m in diameter may appear on the surface. On a weakly stable area, where karst sulfate rocks are overlain by eluvial-diluvial and karst deposits up to 25 m thick, there are conditions for the formation of large underground cavities, and on the surface, there are sinkholes with a diameter of up to 12 m. Unstable areas are characterized by a high occurrence of gypsum and anhydrite relative to the groundwater level and the absence of cover formations. Very unstable areas are confined to the high karst slopes of the Yasyl River and dry valleys.

For more than a hundred years of observations by the Kungur laboratory of the Mining Institute of the Ural Branch of the Russian Academy of Sciences, 167 karst sinkholes have been recorded in the Irensky karst region. In the

Fig. 3.130 Lake Shchuchye after water absorption in the sinkhole at the bottom in spring 2019 (*photo by* O. Kadebskaya)

Fig. 3.131 Karst sinkhole in the Orda cave area (*photo by* N. Maksimovich)

Shlyapniki village, from 1956 to 1978, six sinkholes were recorded, of which the largest (1972) with a diameter of 30 m was accompanied by a light earthquake.

In the Krasny Yasyl village from 1959 to 1979, there were seven sinkholes, including 10 by 12 m in size, 8 m deep near the hostel in 1977; in 1979, there was a sinkhole in the granary. In 1949, on the southwestern outskirts of the Opachevka village, a local resident collapsed and drowned in an opened underground reservoir. Not far from the village in the 60 s, a caterpillar tractor collapsed along with the ground in a field between Klimikha and Gubany during night plowing. In 1905, in the Ozerki village, a peasant house with all the buildings collapsed, and a small lake soon appeared in the hole.

Caves. In the reference book (Caves of the Urals ... 1992), 64 karst caves with a length of 5 m and more are noted within the region. Over the past decades, several more have been added to the indicated number, but, unfortunately, some of the caves have become inaccessible for visiting. There were 86 caves and grottoes registered in the Kungur laboratory of the Mining Institute of the Ural Branch of the Russian Academy of Sciences and the archives of the Association of Cavers of the Urals in the Irensky karst region. The largest number of underground karst forms is confined to rocky outcrops of sulfate rocks on the sides of river valleys (70%) or ravines (30%). At the same time, the entrances of the studied caves are often located at the base of rock massifs (75%), less often in their slope (20%) or watershed parts (20%). Caves are rapidly evolving, changing their size and configuration (Gerasimova et al. 2019).

Here is a description of the largest and deepest caves in the region.

The Orda Cave. The entrance is located in a funnel on the southern slope of Kazakovskaya mountain on the left bank of the Kungur River near the southwestern outskirts of the Orda village. The entrance was opened as a result of a hole with a diameter of about 35 m.

Previously, it was believed that the cave had long been known to local residents (Maksimovich et al. 2006). However, most likely the cave became accessible in the late 1980s as a result of a sinkhole. On Kazakovskaya mountain in 1986, a party under the leadership of N. A. Darovskikh worked on a detailed exploration of fluorite for use as a flux. In the materials of the reports and according to the recollections of the party employees, there was no entrance to the cave at that time. Kazakovskaya Cave, 10 m long, mentioned in the work of G. A. Maksimovich (1969), later renamed Ryabinovaya, was discovered later (Kadebskaya et al. 2009). Exploration of the Orda Cave began in 1992; therefore, the sinkhole and the entrance formed between 1986 and 1992.

It is the longest underwater sulfate cave in the world and the longest underwater gallery system in Russia. The characteristics of the geological, hydrogeological, geomorphological, geochemical, and other conditions of the cave area, as well as its underwater part, are presented in numerous publications (Kadebskaya and Maksimovich 2009, 2012, 2016, 2017; Kadebskaya et al. 2009, 2019a, b; Lyagushkin et al. 2011; Maksimovich 2008; Maksimovich et al. 2006; Maximovich and Shumilova 2008).

The cave was laid in the carbonate-sulfate strata of the Irensky horizon of the Kungurian stage of the Permian system. Conventionally, the cave can be divided into a dry part and a system of underwater galleries with dry grottoes. The total length according to the survey data is 4900 m, including the underwater part—4600 m (Fig. 3.132). According to the 2019 semi-instrumental topographic survey, the length of the main passages was 305 m, and the amplitude was 29 m (Fig. 3.133).

The dry part consists of several grottoes connected by corridors or narrow passages developed at different levels along the systems of cracks of different genesis.

In winter, seasonal glaciation forms in the dry part of the cave in stalagmites, stalactites, and sublimation crystals. The chemical composition of the ice is close to the composition of the surface and ground waters of the cave area and is characterized by a pronounced sulfate composition (Maksimovich 2008).

The underwater part of the cave is the longest siphon recorded in the caves of Russia (940 m). Analysis of the video footage of underwater shooting performed by A. G. Filimonov and A. A. Gorbunov revealed several depressions of a circular cross-section up to 1 m deep at the bottom of the cave (Fig. 3.134), which are compared with ascending subaquatic sources (feeders in the terminology of A. B. Klimchuk). The dimensions of these depressions significantly increased over the course of 3–4 years, and during a flood in the grottoes of the Great Hall, Basement, in the Main Gallery, the Canyon, the Krasnoyarsk and Chelyabinsk passages, pulsating jets emerge from them (Kadebskaya and Maksimovich 2016).

Groundwater in this area belongs to the sulfate-hydrocarbonate-calcium hydrochemical facies and has a fairly high mineralization (up to 2400 mg/dm^3), typical for areas where sulfate karst develops (Maksimovich et al. 2006).

Different directions of groundwater movement were recorded in the cave. The vertical movement of water was noted by A. A. Gorbunov during the spring flood of 2009. Water with a suspension of clay particles moved from the lower horizontal gallery of the Basement passage along an almost vertical channel of the Canyon passage to the overlying horizontal gallery of the Main passage. The upward movement of water is also confirmed by the internal morphology of the cave: at the intersection of the Canyon gallery and the Main gallery on the vault, there is a so-called Big bubble (dome) with a diameter of about 12 m. Several smaller domes can be traced on the ceiling of the Main passage and towards the Great Hall (Fig. 3.135).

Another direction of water movement was noted in the Moskovsky and Sverdlovsky passages, the origin of which is associated with the development of near-side resistance cracks, where they are consistent with the direction of the flow of the Kungur River. The high fracturing of the karst massif facilitates the penetration of turbid waters during a flood from the river system into the distant part of the galleries. Flood waters reach the dry part of the cave in about 5–7 days. Murky water does not enter the Krasnoyarsky and Chelyabinsky passages. Even during low water periods, when suspension rises from the bottom of the cave in the Moskovsky passage, then it moves to the dry part, which indicates the movement of water in a northeastern direction (Kadebskaya and Maksimovich 2016).

Tectonic fragmentation of the massif, caused by the presence of a diagonal system of faults and associated with faults of northeastern strike, determined the configuration of the cave and the morphology of its main galleries. The cave, traditionally considered as the result of downward

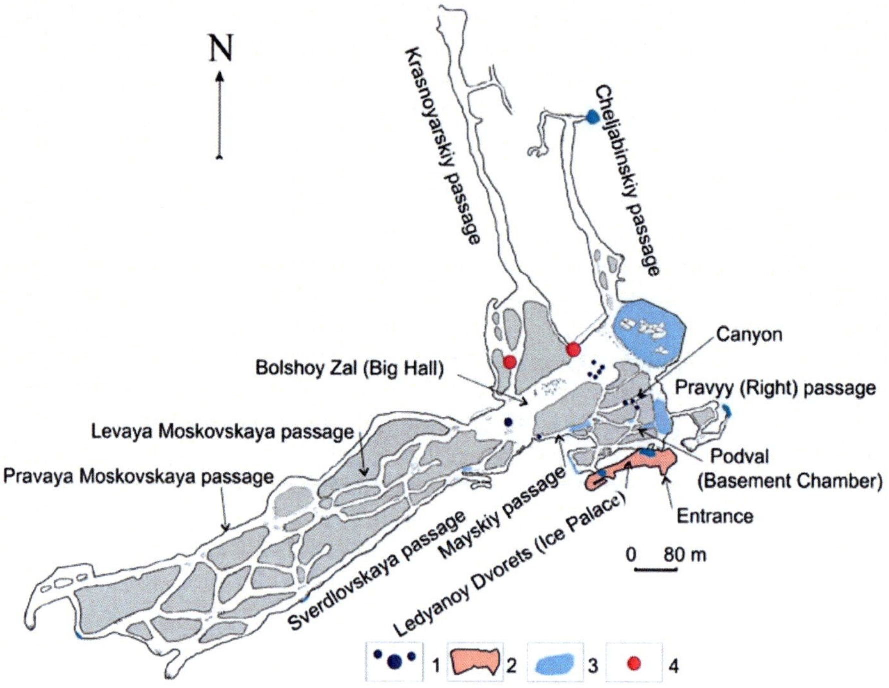

Fig. 3.132 Plan of the Orda Cave (compiled by Osipov D. V. in January 2010 and updated by Zhigalov A. in 2016, with permission): 1—subaqueous springs (feeders), 2—dry part of the cave with lakes, 3—air cupolas in the underwater part of cave, 4—vertical pipes

infiltration, is associated with a deep inflow of a tectonically exposed confined horizon. The ascending springs revealed in the Orda Cave, characterized by a relatively high content of hydrocarbons, suggest that the confined waters come from limestones and dolomites of the Philippovsky horizon. This aquifer is fed on the Ufa plateau. Neotectonic activation, which caused the vertical movement of the blocks, led to the drainage of these pressure waters and the further formation of the cave. According to the classification of Klimchouk et al. (2017), such caves are hypogenic (Kadebskaya and Maksimovich 2016; Kadebskaya et al. 2017, 2019a, b; Maksimovich et al. 2006).

Examination of the vicinity of the cave showed that it may have been of greater size (Fig. 3.136). In the northern direction from the Chelyabinsky and Krasnoyarsky passages, there is a karst ravine with a length of about 3.5 km. The bottom and sides of the ravine are covered with sinkholes of collapse origin. The sinkholes actually go one into the other (Fig. 3.137).

Fresh sinkholes are noted here. The ravine ends with a powerful Podzuevsky spring (Fig. 3.128). A similar ravine of the same direction, with a length of about 2.5 km, is located south of the Arsenovsky spring, on the opposite bank of the Kungur River from the entrance to the Orda cave (Fig. 3.129). It can be assumed with a high degree of certainty that a powerful cave system existed in this area before the formation of the sinkholes.

The manifestation of unusual boric mineralization (howlite), established in the underwater part of the Orda Cave, puts it in the category of unique objects. It is interesting that this mineral was first discovered in underwater photographs taken by M. Yusupova (Maksimovich et al. 2006), and then it was selected for research.

The cave is also inhabited by the relict Khlebnikov amphipod (Crangonyx chlebnikovi).

The Nizhnemikhailovskaya Cave (Niznemikhailovskaya-1, Niznemikhaylovskaya-2) is located 2.5 km southwest of the outskirts of the Vtorye Klyuchiki village, on the left bank of the Iren River in an array of gypsum anhydrites, forming rocky outcrops up to 10 m high.

The cave has three entrances located at a height of 2.5–3.5 m from the edge of the old riverbed of the Iren River. It is a one-story horizontal system of low labyrinth passages developed along cracks mainly in the northwest direction (Fig. 3.138).

The cave contains niches, gutters and other forms of leaching, clay deposits with hummocks and drying cracks, and talus falls under organ pipes. In cold weather, ice formations and sublimation crystals form in the entrance area.

The development of the cave under the conditions of annual fluctuations in the level and flooding from the side of the river is indicated by the characteristic sections of passages with angular walls and even horizontal arches and specific deposits on the bottoms, in niches and pockets of the walls (Andreychuk and Yatsyna 1992). Non-freezing ravines in the river bed are evidence of unloading of water streams.

According to the data of the semi-instrumental topographic survey in 2018, the length of the main passages was

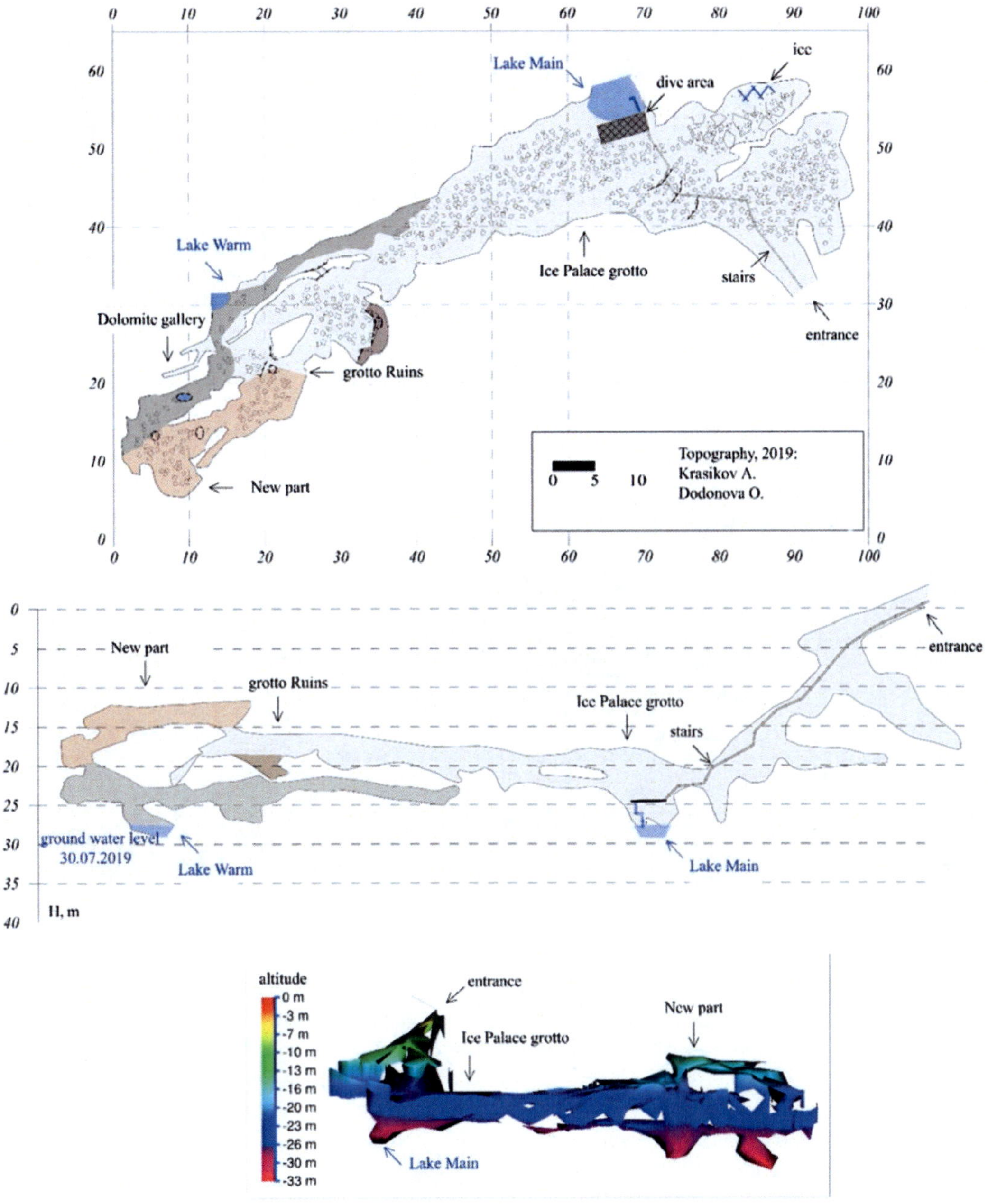

Fig. 3.133 Dry part of the Orda Cave (Krasikov, Dodonova 2019)

1300 m, and the amplitude was 5 m. According to the plan drawn up by the staff of the Kungur laboratory of the Mining Institute of the Ural Branch of the Russian Academy of Sciences in 1989, in the far part of the cave, there is a section with a length of at least 400 m, the passage to which at the time of research (December 2018) was blocked by a clay dump.

The Kashinskaya Cave (Uinskaya 3) is located 3.5 km northeast of the outskirts of the Uinskoye village, downstream of the Aspa River, on the slope of the left bank valley, at the base of a low rocky outcrop of gypsum anhydrites (Fig. 3.139).

An inclined descent, starting from the entrance, leads to the central grotto 15 m long and up to 8 m wide. The main passage to the far part of the cave, which is a labyrinth of corridors and grottoes with avalanche deposits, is located in the western part of the central grotto. The labyrinthine part is developed along cracks, at the intersections of which there

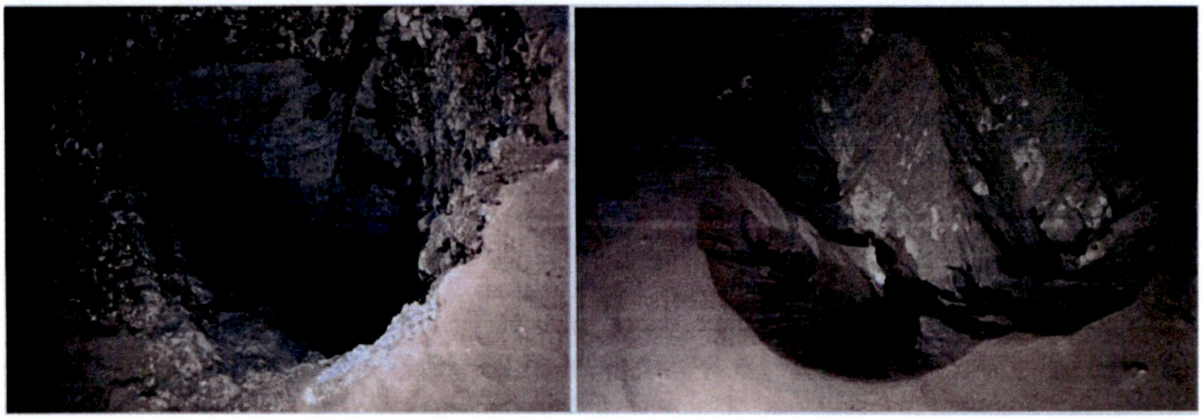

Fig. 3.134 Subaqueous sources (feeders) in the Orda Cave, which control the discharge of hydrocarbonate-calcium waters (*photo by* A. Gorbunov)

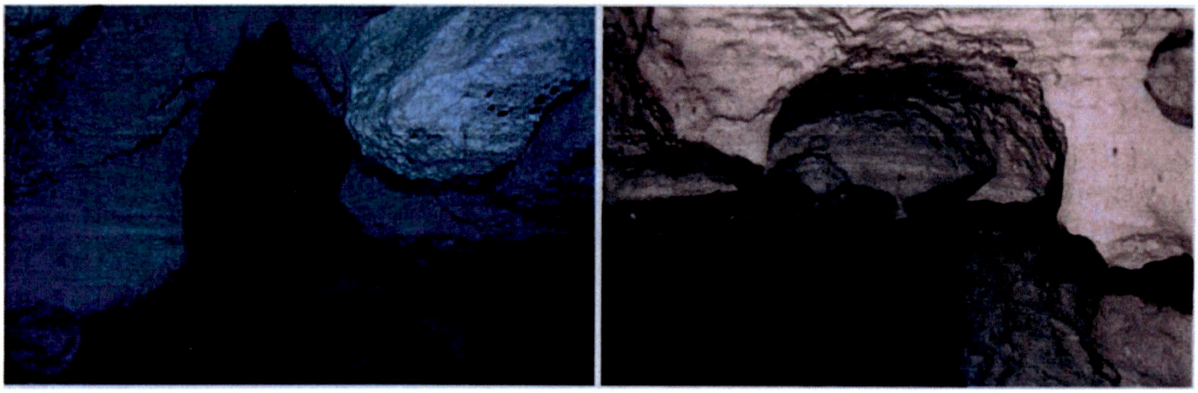

Fig. 3.135 Ellipsoidal domes in the canyon course and spherical domes on the ceiling of the main passage (*photo by* A. Gorbunov)

are several grottoes up to 6 m high, gradually rising to the surface of the earth. Most of the grottoes are interconnected by narrow passages at different levels, forming an extensive system.

The cave is multi-story, corridor-grotto, labyrinth with an underground lake. Landslides, residual, and water mechanical deposits are found. According to the 2019 semi-instrumental survey, the length of the dry part of the cave was 210 m, and the amplitude was 8 m. The watered part does not exceed 10 m and has a depth of 2 m (Fig. 3.140).

The Ponomarevskaya Cave is located 3 km northwest of the Opachevka village in the upper reaches of the left slope of the Yasyl log. The cave is one-story, horizontal, corridor grotto (Kilin and Minkevich 1999) with short impassable lateral branches. The entrance in the form of an arch measuring 6.2 by 1.3 m is located at the base of the gypsum anhydrite outlet at the southern end of the landslide-karst ditch. A wide tunnel with a predominant direction of 60–70° leads from the entrance, crossed by a stream (Fig. 3.141).

According to special studies (Kataev 1999), cave streams represent a fragment of a localized groundwater runoff route. During the flood period, the galleries of the cave are possibly filled with water, as evidenced by the corrosion-erosion shelves in the middle parts of the walls. In low-water periods, residual micro-lakes are observed in the bottom depressions.

The cave contains landslides, residual and water mechanical sediments. In the entrance grotto, seasonal ice stalagmites, and icing crust are observed.

According to the data of the semi-instrumental survey in 2019, the length of the dry part of the cave was 160 m. Currently, the cave is polluted with oil products; in the air, there is a characteristic smell of hydrogen sulfide.

The Sudinsky Proval Cave (Underground Aquapark, Kulakovskaya 3) is located on the right bank of the Sudinka River, 1.6 km east of its confluence with the Iren River, at a distance of 380 m downstream from the absorption site of the river (the Rock Karavashek/Karavay) and 420 m—from

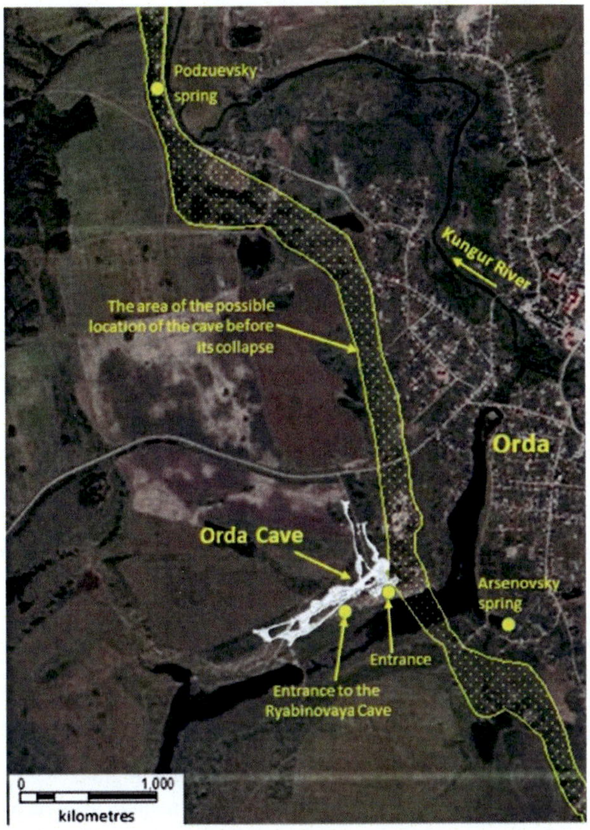

Fig. 3.136 A diagram of the possible location of the Orda Cave before its collapse

the outlet of the river to the surface (Fig. 3.142). It is an underground bed of the Sudinka River.

The triangular entrance of 1.2 by 1.5 m is located in a dip in the slope of the 6.5 m massif at a height of about 15 m at the level of the river valley. A steep descent along an inclined course with a vertical step 2 m high with a total height of 14 m ends with a small platform with a lake, which is a siphon.

To the east of the lake, there is a wide meandering gallery going upstream of the river (Figs. 3.143 and 3.144).

According to the 2019 semi-instrumental survey, the length of the dry part of the cave was 150 m, and the amplitude was 16 m.

The Uinskaya Ledyanaya (Dead Man's) Cave is located 3 km northeast of the eastern border of the Uinskoye village, in a mass of gypsum anhydrites on the right side of the valley at a distance of 0.9 km from the Aspa River. The 0.4 × 0.5 m entrance is located at the base of a 3–4 m high gypsum anhydrite outcrop at the foot of Chaikinsky Mys mountain, 4.6 km below the Uinsky village.

It is a system of low (0.5–1.2 m) passages developed along cracks of various genesis, mainly in the southeast direction (Figs. 3.145 and 3.146). At the intersections of the passages, there are low grottoes up to 1.5 m high and 2–5 m wide.

The cave was excavated in the gypsum of the Irensky horizon of the Kungurian stage. The cave is one-story horizontal slit-grotto, developed along cracks in the meridional and northeastern directions. In addition to landslide-blocky, water mechanical deposits are observed in the cave: clay and sand. In cold weather, ice formations and sublimation crystals form in the entrance part.

The cave was mentioned by V. A. Varsanofieva back in 1915, and, in 1963, L. A. Shimanovsky carried out a detailed study of the cave; 176 m of passages were mapped and the chemical composition of ice and snow was studied. In 1967, the length of the cave increased to 220 m according to the results of the survey carried out by G. N. Panarina. One of the passages at the time of research (March 2019) was blocked by a local clay-clastic-rubble fallout.

The Zmeinaya Cave (Uinskaya-2). Located near the Uinskaya Ice Cave, on the right side of the valley, at a

Fig. 3.137 Karst ravine with numerous sinkholes north of the Chelyabinsk and Krasnoyarsk passages of the Orda Cave (*photo by* N. Maksimovich)

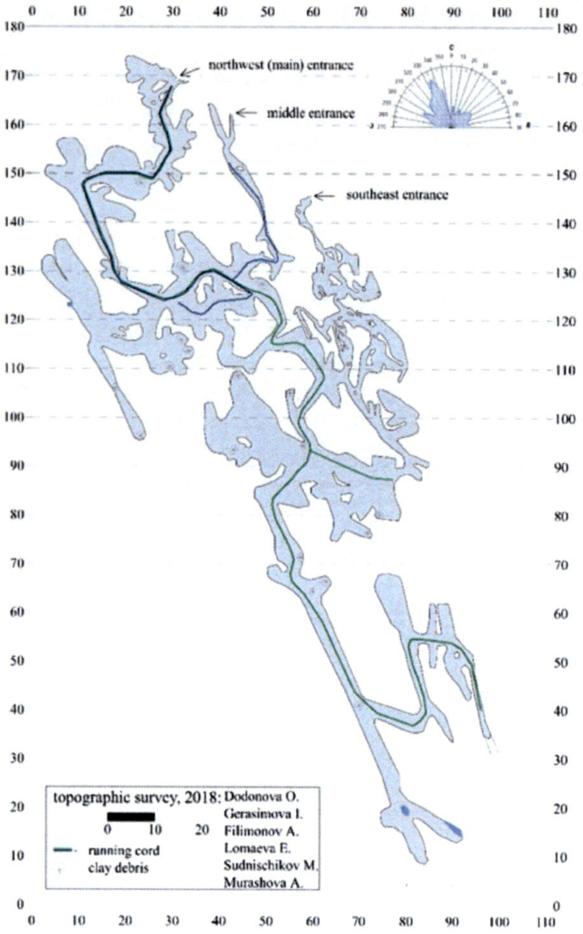

Fig. 3.138 The plan of the Nizhnemikhailovskaya cave (Dodonova, Gerasimova, Filimonov, Lomaeva, Sudnischikov, Murashova 2018)

distance of 0.9 m from the Aspa River. It is located on the territory of the protected natural landscape Uinskoe swamp.

The entrance to the cave is formed at the base of a rocky ledge 7 m high at the foot of Chaikinsky Mys Mountain, 3.5 km below the Uinsky village. The 37 m long cave was excavated in the same rocks as Uinskaya Ledyanaya one. It is located next to Lake Zmeinoye, thanks to which it got its name. The lake water has a smell of hydrogen sulfide.

In March 2008, ice was found in the first grotto along the right wall with a length of 19 m and a width of 0.5–1.5 m. Milky-white ice was covered with a layer of plaster flour.

The Berezovogorskaya Cave is located 1.5 km southwest of the Berezovaya Gora village in a funnel with a diameter of 70 m and a depth of 20 m, complicating the karst ravine. The entrance in the form of a gap up to 0.8 m in diameter is located in a blockage at the bottom of the funnel near the base of a twenty-meter outcrop of gypsum anhydrites.

At the time of the survey (March 2019), in the blockage at the bottom of the grotto, two cracks with a width of 0.5 m are observed, which are locally covered with stone talus. It is a seasonally acting ponor.

According to the 2019 semi-instrumental survey, the length of the main passages was 64 m, and the depth was 20 m (Fig. 3.147).

Existing and proposed specially protected natural areas

Existing SPNAs. Within the Irensky karst region, 13 protected areas of regional and local significance with a total area of 185.5 ha have been allocated for protection (Specially Protected … 2017).

The **Altynnovskaya** forest-steppe is a natural reserve of local importance with an area of 1 hectare, located 3 km southwest of the Altynnoye village. The Chakra-Tau mountain is a natural reserve of local importance with an area of 2.3 ha, located south-west of the Bikbay village. Proposed for protection in 2010. Typical ecosystems of the Kungur forest-steppe grow on the rocky outcrops of sulfate rocks.

The **Duvansky ravine** is a natural reserve of local importance with an area of 10.8 ha, located 600 m east of the Shatunova village. Typical karst ravine with meadow steppe vegetation, characteristic of the Irensky karst region.

The **Ishimovskaya forest-steppe** is a natural reserve of local importance with an area of 23.7 ha, located on the eastern outskirts of the Ishimovo village. It was proposed for protection in 2010. The nature reserve consists of 2 areas: within the northern area, a meadow steppe has formed, and, in the south, a mixed forb forest is located.

The **Lysaya mountain** is a landscape natural monument of regional significance with an area of 37 ha. It is located in rocky outcrops on the right bank of the Iren River. A small Lysaya Gora Cave, no more than 10 m long, is known here. 14 plants listed in the Red Book of the Perm Region were found here.

Lake Samokhvalovo of karst origin is a natural reserve of local importance with an area of 3.4 ha, located 3 km northwest of the Bogorodsk village, 2 km southeast of the Zuevsky village in the forest. Proposed for protection in 2010. The area of the lake is 0.8 ha.

The **Orda Cave** is located within the Kazakovskaya mountain, 2 km west of the Orda village, and is a geological monument of regional significance with an area of 36.6 ha. It is the longest underwater gypsum cave in the world (see description above).

The **Uinskaya Ledyanaya (Dead Man's) Cave** is located 3 km northeast of the eastern border of the Uinskoye village, in an array of gypsum anhydrites on the territory that is part of the regional protected natural landscape Uinskoye

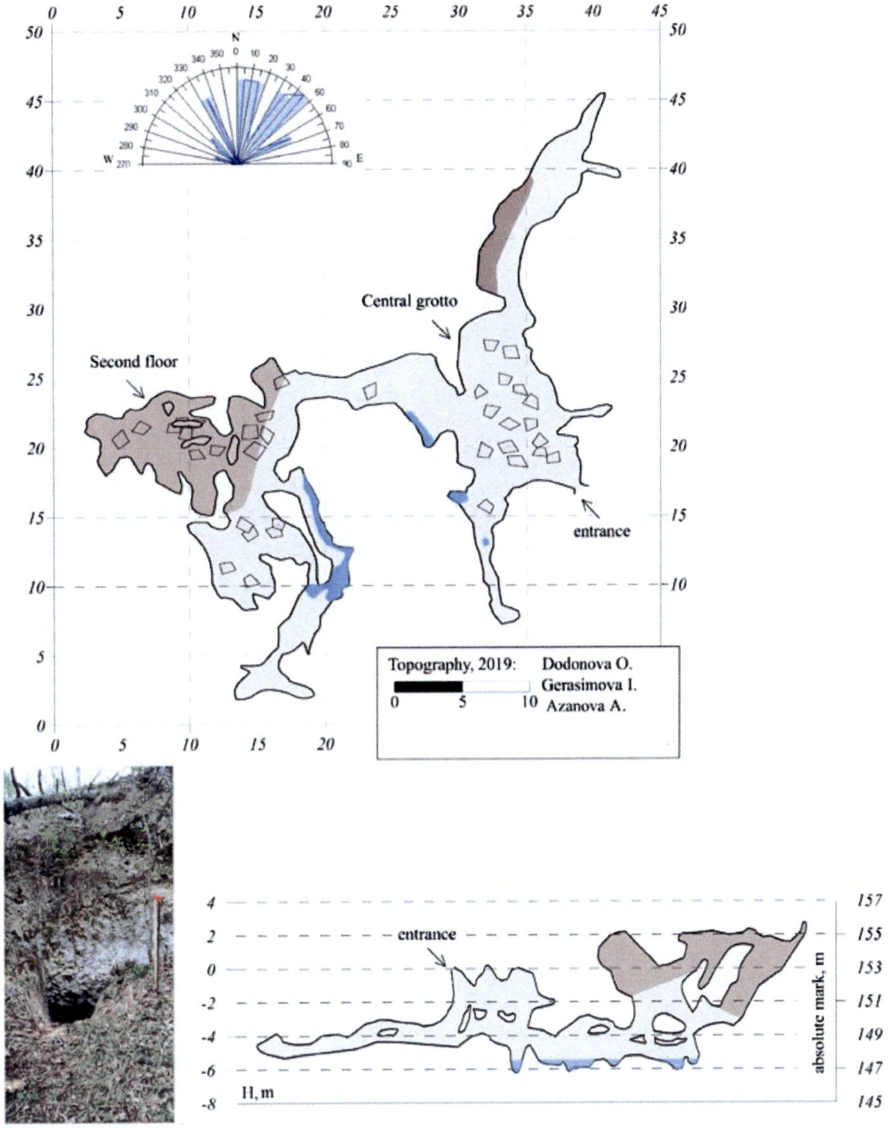

Fig. 3.139 The Kashinskaya Cave plan (Dodonova, Gerasimova, Azanova 2019)

swamp, and at the same time is a geological natural monument of local importance with an area of 0.5 ha.

The **network of lakes Turaevka, Zmeevka, and Orlovka**, located 3 km south of the Novopetrovka village, in the valley of the disappearing karst Tesyak River. Proposed for protection in 2010. They are a natural reserve of local importance with an area of 13 ha. The hydrological features of the lakes are poorly studied.

The **Lachin-Tash Rock** is a natural reserve of local importance with an area of 1.5 ha, located 1.5 km east of the Urazmatyevo village. A gypsum outcrop with forest-steppe vegetation is located on the right side of a karst ravine, the mouth of which goes to the bank of the Iren River.

Lake Tyuynoye of karst origin is a natural reserve of local importance with an area of 37.7 ha, located 2 km east of the Tyuisk village. The total area of the water surface is 22.0 ha. Hydrogen sulfide sources are unloaded on the eastern shore of the lake. There is information about the healing properties of the silt that composes the bottom of the lake. Thus, according to A. A. Oborin and A. R. Ageeva, these therapeutic muds are similar in their properties to those of the Suksunsky pond (see Sect. 3.7).

Lake Chayechnoye is a landscape natural monument with an area of 14 ha. Located in a sinkhole southwest of the Voskresenskoye village. Proposed by A. I. Shurakov in 1981 for the protection of one of the largest colonies of black-headed gulls. The lake is in the stage of waterlogging, and there is drift on it.

Yama Millionnaya is one of the largest sinkholes in the area and a natural reserve of local importance, which is

Fig. 3.140 Survey of the underground lake in the Kashinskaya Cave by D. Mikhalev (*photo by* M. Zhigulina)

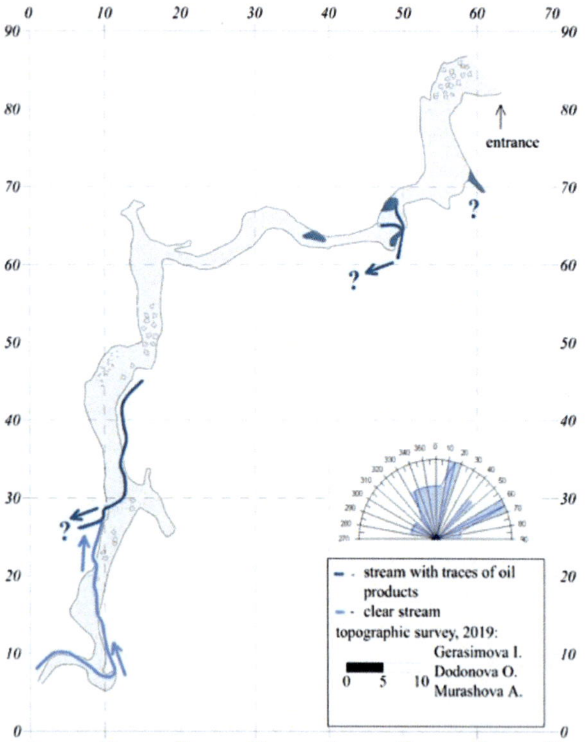

Fig. 3.141 Plan of the Ponomarevskaya Cave (Gerasimova, Dodonova, Murashova 2019)

located 2.5 km northwest of the Tyinsk village. The karst field where Yama Millionnaya is located has an area of about 4 ha, within which more than 100 surface karst forms have been recorded. At the bottom of the funnel, there is a ponor absorbing surface water and a 9 m long cave.

Proposed SPNAs. The most promising object for the creation of protected areas is the site where the **underground bed of the Sudinka River**. On the proposed site, one can observe the absorption of the river and its exit from the ground after 420 m. The underground watercourse itself can be traced in the cave, which is a unique phenomenon even for such a well-studied karst region as Irensky, with a wide variety of karst forms.

Economic activity and environmental pressure. The main production profile in the territory of the Irensky karst region is agricultural. Extraction of oil, building and finishing stone, sand, clay, limestone, and gypsum is carried out.

Influence of oil fields. Oil fields being developed: Preobrazhenskoye, Sharonovskoye, Sofronitskoye, Dozortsevskoye, Dulepovskoye, Kokuyskoye, and Churakovskoye. Commercial oil production began in 1952 and continues to the present day.

One of the unique underground rivers—Melenka—is known within the Churakovskoye oil field. Its underground channel was opened in a large karst gap with a diameter of

Fig. 3.142 The Rock Karavay and the place where the Sudinka River goes underground (*photo by* M. Zhigulina)

Fig. 3.143 Underground river bed in the Sudinsky Proval cave (*photo by* O. Kadebskaya)

about 100 m. The river is polluted with oil products, now an oil trap is installed at the place of the sinkhole (Fig. 3.148).

At the Kokuyskoye oil field, two contaminated areas are distinguished: on the Kamenka River, with chloride content in the waters up to 840 mg/dm^3, and on the Pavlovka River (the former Turaevka River), where the waters have a smell of hydrogen sulfide and a whitish color (Fig. 3.149). Some of their channels are underground, unloading occurs in the form of karst springs.

Until the 90 s of the twentieth century, the water in the rivers was clean and used for drinking. In April 1997, a volley of oil was released into a large karst spring feeding the Pavlovka River.

According to the 1996 sampling data, the chloride content in the Pavlovka River is 906 mg/dm^3, and in the Kamenka River—836, 477, and 488 mg/dm^3, respectively, from the source to the mouth. In the spring located between the Zakharovka and Pavlovo villages, there is an increased content of chlorides up to 1055 mg/dm^3. The state of groundwater according to the sampling of springs has improved. The chloride content in the waters at the Kamenka River estuary and in the Pavlovka River basin is also decreasing.

According to the results of sampling in 1997 in the Kamenka River and nearby springs (except for the source), an increased content of phenol is noted with an excess of MPC by 5–6 times, in the area of the Zakharovka and

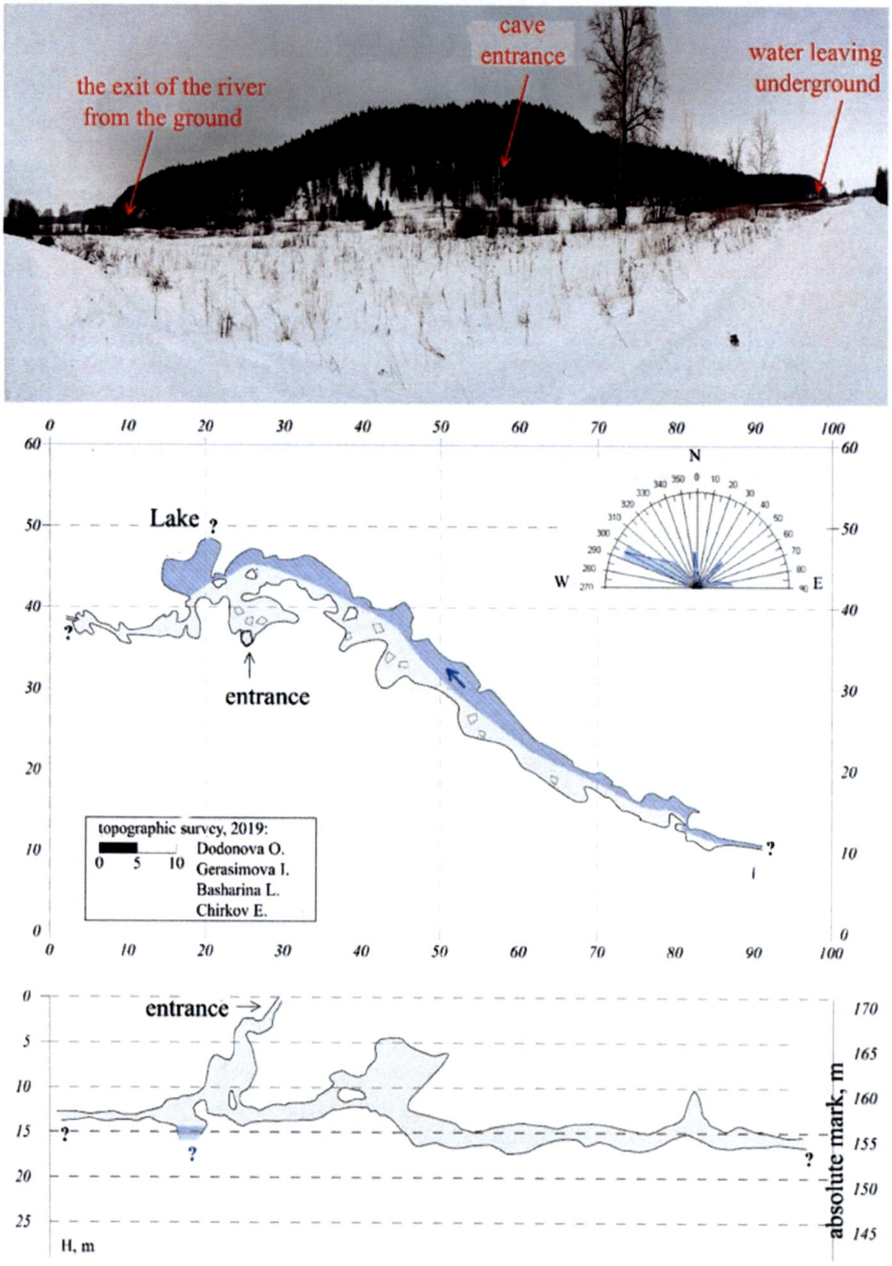

Fig. 3.144 Plan of the Sudinsky Proval Cave (Dodonova, Gerasimova, Basharina, Chirkov 2019)

Pavlovo villages—up to 0.005 mg/dm^3. The content of oil products in the head of the Kamenka River is 0.2 mg/dm^3, and, in the Pavlovka River, it varies from 1.5 to 120 mg/dm^3, amounting to 6 mg/dm^3 at the mouth (Yashchenko 1984).

Currently, the water of the Pavlovka and Kamenka rivers contains saturated and unsaturated hydrocarbons, their acids, ethers, alcohols, and halogenated derivatives (Fig. 3.150).

The bottom of the rivers is covered with a thin layer of dark brown oil tar a few tens of meters below their outcrops from the rocks, and subaqueous springs with oil are periodically unloaded in the river beds. River waters are characterized by a strong smell of hydrogen sulfide emanating from them (Fig. 3.151).

The development of a bacterial sulfate-reducing process is caused by the ingress of organic matter (oil products) into waters with a high sulfate content:

$$2SO_4 + 5C_{org} + 2H_2O = 2H_2S + 5CO_2,$$

where petroleum hydrocarbons serve as carbon and energy sources.

Fig. 3.145 Slump-block deposits in the grottoes of the Uinskaya Ledyanaya Cave (*photo by* M. Ovseichik)

Another product of the bacterial sulfate reduction process is carbon dioxide. Its entry into groundwater contributes to the precipitation of secondary calcite, and the waters acquire a whitish color:

$$Ca + CO_2 + H_2O = CaCO_3 + 2H.$$

This factor reduces the saturation of groundwater with calcium ions and also contributes to the further dissolution of rocks. Bacterial recovery of sulfates from petroleum hydrocarbons is accompanied by the enrichment of karst waters with organic acids. Bacterial oxidation of oil hydrocarbons under aerobic conditions also leads to an increase in the content of organic acids in water, which dissolve sulfate rocks.

Thus, oil pollution leads to an intensification of karst processes in both aerobic and anaerobic conditions (Maksimovich et al. 2016a, b).

Pipeline transport. The karst areas between the Kungur and Iren rivers are crossed by the routes of the Siberia-Center-West gas pipelines. Here, the gypsum and anhydrite of the Lunezhsky and Demidkovskaya suite located between them, are the most karst in the upper (up to a depth of 40 m) part of the section. In the zone of main gas pipelines, the activation of karst is observed, caused by the disturbance of the cover sediments and the redistribution of the surface runoff. The most karsted areas are where gypsum comes to the surface, covered with a soil-vegetation layer or eluvium of low thickness. The number of sinkholes here reaches 95 forms/ha, their total area occupies about 50% of the site, and the average conditional decrease in the surface due to sinkholes is 57 cm. The initial dimensions of the sinkholes are 2–3 m, and the average diameter of the formed sinkholes is 7–8 m. Since May 1983 to October 1984, 24 dips appeared in the strip of the gas pipeline 40 m wide and 5.4 km long, and in 1985 their number exceeded 45. The diameter of most of the dips is not more than 2.5 m, the depth is 2 m, and only, in some cases, 5 m (Gorbunova and Maksimovich 1991; Kostarev et al. 1986).

Karst processes in the area of the main gas pipeline routes were studied in the Yasylsky karst site (Kilin and Minkevich 2018), where sinkholes of different genesis were formed: corrosion-sinkhole, corrosion-suffosion, corrosion-subsidence, corrosion-erosion, sinkhole, and polygenetic. After the construction of the gas pipeline, activation of karst was noted in the ravines with the greatest distribution of cone-shaped, saucer-shaped funnels, with gently sloping turf walls, with an active bottom that filters well water, mainly with a depth of 1–3 m and more (68–70%). The dimensions in the plan are from 1 by 1 to 20 by 20 m, less often—13 by 40 m (Table 3.7).

As anti-karst measures, the method of tamping gas pipelines with clay soils is used. This method is used when

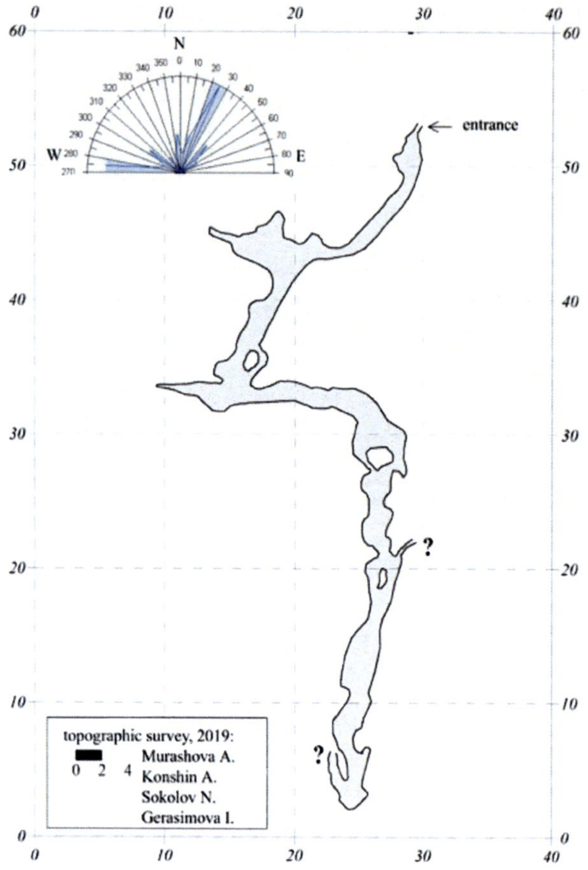

Fig. 3.146 Plan of the Uinskaya Ledyanaya Cave (Murashova, Konshin, Sokolov, Gerasimova 2019)

the stress on the pipes, according to technical monitoring data, reaches critical values. The track is dug up and the soil is tamped under the base, thereby unloading the pipe from the critical values of the maximum permissible level of metal stresses. In places of large karst cavities and in active zones, the pipe is hung on bored piles, which are buried below the karst rocks. Surface karst sinkholes, ponors are plugged with clay using geogrids, and cement plugs (Kilin and Kilina 2018).

Railway lines. The Kazan–Yekaterinburg railway runs in the south of the region. Here, under the newly built track at the Ater junction, on September 24, 1965, due to the lack of drainage, a 3-m-in-diameter sinkhole arose. In the same area, 2.5 m from the tracks, on July 15, 1961, a 9.5 m deep sinkhole with a diameter of 5 m was formed Subsequently, on the site of the filled-in depression, subsidence of the earth's surface was noted for a number of years (Yakovenko 1968). The formation of sinkhole suspended the movement of trains for a long time or forced to reduce the speed of their movement to 15–20 km/h.

Plaster handicraft. Irensky karst region is the richest in reserves, including ornamental, gypsum in the region. The most famous deposits are located on the site of the Pavlovo village—the Krasny Yasyl village—the Opachevka village (Fedorovskoe, Bogomolovskoe, Andreevskoe, Denisovskoe, Na Shumu). Gray and brown gypsum is known at the Churakovskoye and Yakovlevskoye deposits; colored—at Zavodskaya Lesosseka, Raspilovochnaya, white—at Kopantsy, Grishinsky, Zakharovsky-1, Pologiy log, Yasylsky logs, Pavlovsky, Burovoy, Egorshiny, and Levushkin Yamy, Opachevsky, etc. (Geological monuments … 2009).

The gypsum handicraft originated here due to the variety of gypsum types: snow-white, pinkish-gray, spotted, reticulated, yellow, and bluish-gray (Fig. 3.152).

The largest are the following commercial deposits (Darovskikh and Kudryashov 2001).

The Fedorovskoe (Orlovskoe) deposit of selenite and white and brown ornamental gypsum is located on the right bank of the Iren River, 5.5 km north of the Krasny Yasyl village, near the Fedorovka village. The deposit has sinkholes, but the underground karst is poorly developed. The field is not watered.

The Na Shumu deposit of selenite and brown gypsum is located 6 km north of the Krasny Yasyl village (0.4 km north of the Fedorov deposit) (Fig. 3.153). Its dimensions are 300 by 150 m. The area of the deposit is weakly karst (karst coefficient—0.098). The field is not watered.

The Denisovskoe selenite and brown gypsum deposit are located on the right bank of the Iren River, 1.5 km south of the Andreevka village. Its dimensions are 700 by 180 m. The deposit is weakly karst (0.03 forms/km^2) and is not watered.

The Odinovskoye deposit of white ornamental gypsum, 400 by 90 m in size, is located 3.5 km southeast of the Klyuchiki-2 village. The karst content of the deposit is 27%; numerous sinkholes are developed on the surface. Karst breccia is developed near the sublatitudinal fault. Surface open karst appears in the form of sinkholes (138 forms with a total area of 8710 m^2). Underground karst appeared at the deposit in the form of surface, covered and deep. The first type is observed in the southern and western parts of the deposit in the form of a karst-erosional ravine (in the south), as well as sinkholes and small kettles under a layer of sandstones (in the west and north). The second type was observed in the form of voids and cavities up to 2–3 m in size, filled with clay and crushed stone of the host rocks. The field is partially watered.

The Sudinskoe Pole deposit of white and colored ornamental gypsum is located 1.5 km southeast of the Klyuchiki-2village. Located on the southwestern slope of the watershed of the Iren and Kungur Rivers. In plan, it

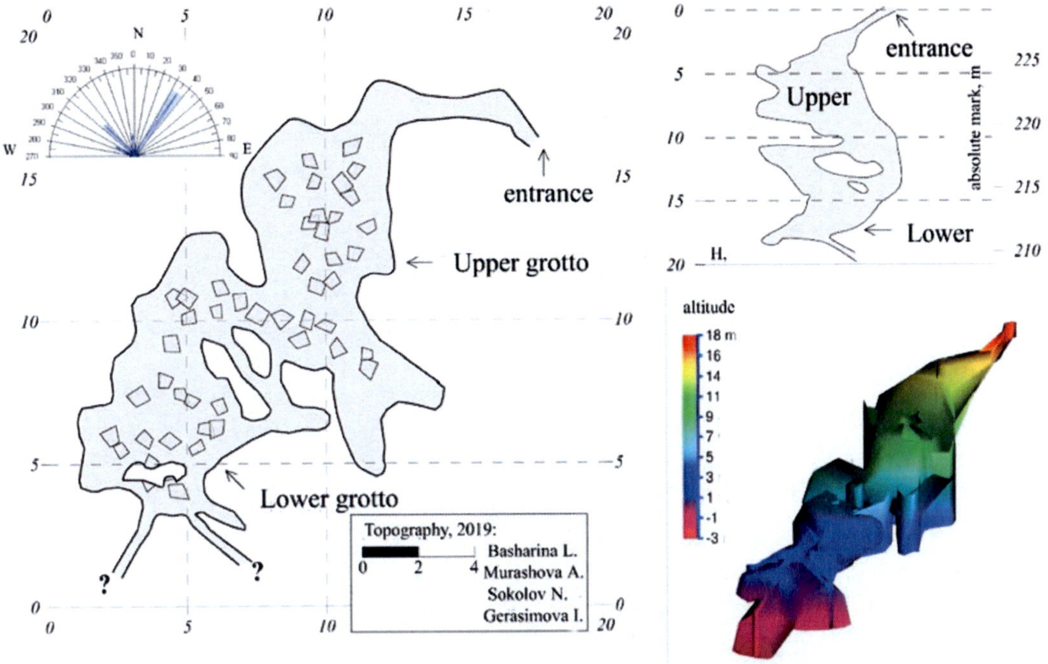

Fig. 3.147 Plan of the Berezovogorskaya Cave (Basharina, Sokolov, Murashova, Gerasimova 2019)

Fig. 3.148 Oil trap on a karst spring within the Churakovskoye field (*photo by* O. Kadebskaya)

Fig. 3.149 Contaminated bed of the Kamenka river within the Kokuyskoye field (*photo by* O. Kadebskaya)

Fig. 3.150 One of the polluted karst springs of the Pavlovka River, equipped with an oil trap (*photo by* O. Kadebskaya)

Fig. 3.151 A spring in the Kamenka River valley with the smell of hydrogen sulfide (*photo by* N. Maksimovich)

Table 3.7 Indicators of surface karst content in the Yasyl karst polygon (Kilin and Minkevich 2018)

Indicator	Years				
	2010	2011	2012	2013	2014
Area, m^2	774,150	774,150	774,150	774,150	774,150
Number of karst forms	28	26	33	31	68
Total area of karst forms, m^2	49.18	35.54	83	308.5	129.53
Total volume of karst forms, m^3	7.04	33.13	135	288.3	151.72
Funnel diameters	0.2–5.0	0.3–3.5	0.3–4	0.8–10	0.1–6
Funnel depth	0.2–3	0.2–2	0.3–3.5	0.8–1.5	0.4–2
Depth indicator	0.2–6.5	0.3–0.5	0.4–3.3	0.44–5.82	0.11–7.5
Density of karst forms	36.2	33.59	42.63	40.05	87.85
Areal coefficient of surface karst content, %	0.006	0.0046	0.010	0.04	0.017
Volumetric indicator of surface karst content, mm	0.009	0.043	0.174	0.025	0.196
Average volume of suffusion-karst sinkhole, m^3	0.25	1.27	4	9.3	2.2

Fig. 3.152 Vases of various varieties of selenite (*photo by* N. Maksimovich)

measures 1000 by 250 m. The deposit is heavily karst (up to 35 sinkholes per 10 thousand m^2) and is partially watered.

The Yakovlevskoe deposit of gray ornamental gypsum is located on the right bank of the Iren River near the Yakovlevka village. 60 karst sinkholes were found at the deposit, up to 10 m deep. The karst content coefficient of the deposit is 0.225. Partially watered.

The Bogomolovskoe gypsum and anhydrite deposit. The deposit showed underground and surface open karst. The issues of the preservation of the gypsum strata are associated with karst processes. Surface-covered karst has the greatest impact on the preservation of the gypsum strata. Areas with relatively high underground karst thickness are located in the central part of the field. The high karst content of 25% is mainly associated with the development of surface-covered karst under the Solikamsky deposits. Open karst is developed throughout the entire area of the deposit. Surface karst appears as sinkholes with a density of 550 forms/km^2. The diameter of the funnels is from 5 to 30 m, on average 16 m, and the depth is from 1.3 to 10.8 m, on average 3.3 m. The funnels are predominantly bowl- and saucer-shaped, covered with vegetation (Alvanyan et al. 2012).

The Kruglovskoye gypsum deposit is located 13.5 km south-southwest of the center of the Kungur town, and 1.9 km east of the Sukhaya Rechka village. A productive

Fig. 3.153 Outputs of ornamental stone (selenite) in the Na Shumu quarry (*photo by* O. Kadebskaya)

gypsum deposit composes the upper part of the Demidkovskaya suite. The occurrence of the layer is horizontal or close to horizontal. The strata of gypsum and anhydrite include interlayers of dolomites, intermediate varieties between them, karst rocks, as well as rare interlayers of marls and sandstones. Underground karst is registered by the holes of the drill string in the voids, as well as by the intervals of clays with gruss and crushed stone of bedrock karst rocks. The thickness of the sinkholes varied from 0.1 to 5.6 m, and their number with a thickness of more than 0.5 m reached 5 cases per well. The thickness of the sinkholes along the useful strata varied from 0.2 to 1.8 m. Filled karst voids were most often observed in the underlying sediments. In the useful stratum, rarely filled voids with a thickness of 0.2 to 2.1 m are noted. Excluding karst intervals, less than 0.5 m thick from the calculation, the karst content of the useful stratum is low and equal to 2.8% (Darovskikh and Suntsev 2013).

References

Actual problems of hydrology and hydrochemistry of the Kama reservoir. (2004). Kitaev, A.B. (ed.). Perm: Perm University. (in Russian).

Agashkov, V.I., & Trushnikova, M.S. (1965). Opokinskie caves // *Peshchery (Caves)*, 5 (6), 102–109. http://nsi.psu.ru/cave/vipuski/v5(6).pdf. (in Russian).

Alvanyan, A.K., Iblaminov, R.G., Konoplev, A.V. (2012). Physical and mechanical properties of gypsum stone and geotechnical conditions of the Bogomolovskoye deposit. *Scientific Journal of KubGAU*, 83 (09). http://ej.kubagro.ru/2012/09/pdf/55.pdf. (in Russian).

Andrejczuk, W. (2007). *Kras jako czynnik geoekologiczny*. Sosnowiec–Symferopol.

Andreychuk, V.N., & Yatsyna, I.I. (1992). Nizhnemikhailovskaya gypsum cave. *Issues of the Ural speleology*, 46–49. (in Russian).

Baryakh, A.A., Devyatkov, S.Yu., Samodelkina, N.A. (2016). Theoretical substantiation of the conditions for the formation of sinkholes on the earth's surface after emergency flooding of potash mines. *Physical and technical problems of the development of minerals*, 1, 50–61. (in Russian).

Baryakh, A.A., Sanfirov, I.A., Diaghilev, R.A. (2013). Monitoring the consequences of flooding a potassium mine. *Gornyi Zhurnal (Mining Journal)*, 6, 34–39. (in Russian).

Belov, Yu.E., Belov, S.Yu., Shestov, I.N. (2004). The main types of mineral medicinal and medicinal-table waters in the area of the sanatorium "Klyuchi". *Actual issues of balneology: materials of the interregion*, 11–14. (in Russian)

Beltyukov, G.V. (1978). Types of karst in the area of the Verkhnekamsky salt-bearing basin. *Karst of marbles, dolomites, reefs, calcareous tuffs and halogen deposits: abstracts*. (in Russian).

Bobrova, V.N. (1960). Some new caves in the Kungurskiy district of the Perm Region and the chemistry of the waters of underground lakes. *Issues of geography and nature protection of the Urals: reports of the 5th All-Ural meeting on the geography and nature protection of the Urals*. 2/4, 1–3. (in Russian).

Bogoslovsky, V.F. (1940). From observations of karst phenomena in the interfluve of the Kama River and the Chusovaya River. *Scientific notes of Molotov Pedagogical University*, 7. (in Russian).

Budanov, N.D. (1964). *Hydrogeology of the Urals*. Moscow: Nauka.

Butyrina, K.G. (1968). *Gypsum karst in the central part of the Perm region: PhD thesis*. (in Russian).

Butyrina, K.G. (1974). Chemical geography of small rivers in the central part of the Perm region. *Hydrogeology and Karst Science*, 5. (in Russian).

Butyrina, K.G. (1975). Karst regions and sites of the central part of the Perm region. *Issues of physical geography of the Urals: scientific notes of Perm University*, 47–65. (in Russian).

Caves of the Volga, Urals and Cis-Urals. (2010). Statistical handbook. Naberezhnye Chelny: NGPI. (in Russian)

Caves of the Urals and Cis-Urals. (1992). compiled by I.A. Lavrov, V. N. Andreychuk. Perm. (in Russian).

Chuvashov, B.I., Dyupina, G.V., Mizens, G.A., Chernykh, V.V. (1990). *Support sections of the Upper Carboniferous and Lower Permian of the western slope of the Urals and Cis-Urals*. Sverdlovsk: Ural Branch of the Academy of Sciences of the USSR. (in Russian).

Darovskikh, N.A. & Kudryashov, A.I. (2001). *Geology and prospecting for ornamental gypsum deposits*. Perm: GI UB RAS. (in Russian).

Darovskikh, N.A., & Suntsev, A.S. (2013). Factors of gypsum content at the Kruglovskoye field in the Perm Territory. *Modern problems of science and education*, 5; URL: www.science-education.ru/111-10461 (date of access: 19.12.2014). (in Russian).

Dorofeev, E.P., & Lukin, V.S. (1969). Subsidence karst troughs in the northern part of the Solikamsk depression. *Issues of karst studies*, 105–107. (in Russian).

Egorov, S.I., Shestov, I.N., Shuvalov, V.M. (2007). Drinking medicinal, medical-table, balneological and industrial underground waters of the Cis-Urals. *Bulletin of Perm University. Geology*, 4 (9), 135–147. (in Russian).

Fetisov, V.V. (2005). *Hydrogeological features of karst rocks in the Middle Urals and Perm Cis-Urals): PhD thesis*. (in Russian).

Galimov, N.R., Gorozhanin, V.M. Kochergin, A.V., Linnik, V.N. (2013). Borate mineralization of the Deikovsky site. *Problems of mineralogy, petrography and metallogeny. Scientific readings in memory of P.N. Chirvinsky*, 16, 62–65. (in Russian).

Geological monuments of the Perm region: encyclopedia. (2009) Tchaikovsky, I.I. (ed.). Perm: Mining Institute of the Ural Branch of the Russian Academy of Sciences. (in Russian).

Gerasimova, I.Yu. (2020). The caves of the Kishertsky speleorion. *Peshchery (Caves)*, 43, 6–13. (in Russian).

Gerasimova, I.Yu., Shvetsova, O.O., Murasheva, A.S. (2019). The caves of the Irensky speleorion. *Peshchery (Caves)*, 42, 16–28. http://nsi.psu.ru/cave/vipuski/v42.pdf. (in Russian).

Golikov, Yu.V. (2017). Diagnostics of the state of a railway embankment and its foundation by a remote electromagnetic method. *Innovative transport*, 4, 34–37. (in Russian).

Gorbunova, K.A. (1954). Karst sinkholes. *Priroda (Nature)*, 12, 111–115. (in Russian).

Gorbunova, K.A. (1957). Geomorphology of the environs of the Ust-Kishert village of Perm region. *Scientific notes of Perm University*, 11(2), 33–49. (in Russian).

Gorbunova, K.A. (1959). Characteristics of the Klyuchevskoy site of the Kishertsko-Suksunsky karst region. *Scientific notes of Perm University*, 14(1), 9–13. (in Russian).

Gorbunova, K.A. (1965). *Features of gypsum karst*. Perm: Perm publishing house. http://nsi.psu.ru/cave/bibl/ogk.pdf. (in Russian).

Gorbunova, K.A. (1977). *Gypsum karst of the USSR*. Perm: Publishing house of Perm University. http://nsi.psu.ru/cave/bibl/kgu.pdf. (in Russian).

Gorbunova, K.A. (1985). *Karst studies. Issues of typology and morphology of karst*. Perm: Perm University5. http://nsi.psu.ru/cave/bibl/kvtimk.pdf. (in Russian).

Gorbunova, K.A., & Maksimovich, N.G. (1981). Types of karst water discharge in areas of gypsum-anhydrite karst. *European Regional Conf. on speleology: collection of materials*, 439–441. http://nsi.psu.ru/labs/gtp/stat/ng_0026.html. (in Russian).

Gorbunova, K.A., & Maksimovich, N.G. (1991). Technogenic impact on the karst territories of the Perm Region. *Geography and natural resources*, 3, 42–46. http://nsi.psu.ru/labs/gtp/stat/ng_0122.html. (in Russian).

Gorbunova, K.A., & Maksimovich, N.G. (1994a). Formation of sulfate-calcic waters in Kungur Cave massif. *Cave and Karst Sciences*, 21(1), 12.

Gorbunova, K.A., & Maksimovich, N.G. (1994b). The Program of Investigation in Kungur Ice Cave. Newsletter. *Geology, Climate, Hydrology and Karst Formation*. Project IGCP, 299, 36.

Gorbunova, K.A., Andreychuk, V.N., Kostarev, V.P., Maksimovich, N. G. (1992). *Karst and caves of the Perm region*. Perm: Publishing house of Perm University. http://nsi.psu.ru/labs/gtp/stat/ng_0129.pdf. (in Russian).

Gorbunova, K.A., Maksimovich, N.G., Blinov, S.M., Kraev, V.G. (1997). Factors determining underground water regime of Kungur Cave (the Urals). *The Proceedings of Karst-Water Environment Symposium*. Blacksburg.

Hydrogeology of the USSR. (1972) Volume XIV. Ural. Preis, V.F. (ed.). Ural Territorial Geological Administration. M.: "Nedra". (in Russian).

Ikonnikov, E.A. (1975). Conditions for the development of karst in the northern part of the Yuryuzan-Sylven depression. *Hydrogeology and Karst Studies*, 6, 51–55. http://nsi.psu.ru/cave/bibl/hik1975.pdf. (in Russian).

Ikonnikov, E.A. et al. (1965). *Report on the results of work on compiling a sheet hydrogeological map of scale* 1:500000 sheets R-39-G, R-40-V, G, O-40-B, Perm. (in Russian).

James, A.N., & Kirkpatric, I.M. (1980). Design of foundations of dams controlling soluble rock and soil. *Quaternary Journal of Engineering Geology and Hydrogeology*, 13, 189–198.

Kadebskaya, O.I., & Maximovich, N.G. (2009). Geological, hydrogeological and hydrogeochemical prerequisites for the formation of the Ordinskaya cave. *Peshchery (Caves)*, 32, 12–21. http://nsi.psu.ru/labs/gtp/stat/2009/0364.pdf. (in Russian).

Kadebskaya, O.I., & Maximovich, N.G. (2012). Scientific research in Orda and Kungur Ice Caves. *Scientific research in show caves: abstracts of international congress Skocjan-Borgo Grotta Gigante*, 27–28.

Kadebskaya, O.I., & Maximovich, N.G. (2016). Ordinskaya Cave as a Geographical Phenomenon of World Significance. *Geographic Bulletin*, 2(37), 17–28. http://nsi.psu.ru/labs/gtp/stat/2016/479.pdf. (in Russian).

Kadebskaya, O.I., & Maximovich, N.G. (2017). The Role of Hypogene Speleogenesis in the Formation of the Ordinskaya Cave, Fore-Urals, Russia. In: Hypogene Karst Regions and Caves of the World: monogr. Switzerland: Springer International Publishing AG. Chapter 26. pp. 431–446.

Kadebskaya, O.I., Maximovich, N.G., Zhakova, U.V. (2009). Karst forms in the Ordinskaya cave area (Perm region, Russia). *Speleology and Karstology*. 3, 60–65. http://nsi.psu.ru/labs/gtp/stat/2009/0365.pdf. (in Russian).

Kadebskaya, O., Kataev, V., Maksimovich, N., Meshcheriakova, O. (2017). Elements of hypogene origin in the karst caves of the Urals (Russia). *Journal of Nepal Geological Society*, 54, 141.

Kadebskaya, O.I., Kataev, V.N., Maksimovich, N.G., Kazantseva, A.S. (2019a). Morphological indications of hypogene speleogenesis in caves of the Urals (Russia). *Environmental Earth Sciences*, 78, 485.

Kadebskaya, O.I., Maksimovich, N.G., Gorbunov, A.A. (2019b). Ordinskaya cave. In: Shelepin, A.L. (ed.) Atlas of caves of Russia. pp. 177–183. Russian Geographical Society, Russian Union of Cavers. Moscow. http://nsi.psu.ru/labs/gtp/stat/2019/541_2019b.pdf . (in Russian).

Kalinina, T.A. (2015) Lithology of the Lower Permian carbonate and sulfate evaporites of the "classical" kungur (Perm Region). *Bulletin of Perm University. Geology*, 3(28), 6–27. (in Russian).

Kalinina, T.A., & Tchaikovsky, I.I. (2014). About the nature of breccia rocks in the top of the salt deposit of the Verkhnekamsky salt deposit of the Cis-Ural foredeep. *Lithology and mineral resoureces*, 5, 424–431. (in Russian).

Kalmykova, M.A., & Dutkevich, G.A. (1934). Geological research in the Northern Urals in 1930–1932. *Proceedings of the Petroleum Geological Prospecting Institute*, A(42), 101. (in Russian)

Kataev, V.N. (1999). Typical small caves of the Irensky karst region. *Peshchery (Caves)*, 25(26), 57–62. http://nsi.psu.ru/cave/vipuski/v25-26.pdf. (in Russian).

Kataev, V.N. (2000). Hydrogeological conditions of the Krasnoyasyl landfill: types of waters, water manifestations and their chemism. *Hydrogeology and Karst Studies*, 13, 127–136. (in Russian).

Kataev, V.N., & Kadebskaya, O.I. (2010). *Geology and karst of the Kungur town*. Perm: Perm State University, Mining Institute of RAS. (in Russian).

Kataev, V.N., Kopantseva, E.N., Ermolovich, I.G. (2017). The modern chemical composition of the waters of the lakes of the junction zone of the Ufa swell and the Ural foredeep. *Bulletin of Perm University. Geology*, 16(4),332–346. (in Russian).

Kazantseva, A.S., Kadebskaya, O.I., Dublyansky, Yu.V., Kataev, V.N. (2020). Hydrogeochemical characteristics of groundwater in the junction zone of the Ufa swell and the Ural trough. *Izvestiya UGGU*, 2(58), 83–96. (in Russian).

Khayrulina, E., & Maksimovich, N. (2018). Influence of drainage with high levels of water-soluble salts on the environment in the Verhnekamskoe potash deposit, Russia. *Mine Water and the Environment*, 37(3), 595–603.

Khmurchik, V.T., & Maksimovich, N.G. (2007). Use of aboriginal microflora to combat oil pollution of groundwater. *Bulletin of Perm University. Biology*, 5(10), 123–126. http://nsi.psu.ru/labs/gtp/stat/2007/0326.pdf. (in Russian).

Khudenkikh, K.O., & Kataev, V.N. (2018). Karst morphogenesis in the near-surface zone of the sulfate-carbonate massif of the Sokolino-Sarkaevsky gypsum and anhydrite deposit. *Bulletin of Perm University. Geology*, 17(2), 171–176. (in Russian).

Kilin, Yu.A., & Kilina, E.Yu. (2018). Methods of karstological research on the routes of main gas pipelines. *Geoecology, engineering geodynamics, geological safety*. Perm. (in Russian).

Kilin, Yu.A., & Minkevich, I.I. (1999). Cavities of the Krasnoyasyl karst field. *Peshchery (Caves)*, 25(26), 52–57. http://nsi.psu.ru/cave/vipuski/v25-26.pdf. (in Russian).

Kilin, Yu.A., & Minkevich, I.I. (2018). New karstological studies in the Perm Kama region. *Issues of geography. Speleology and karst studies*, 147, 17–29. (in Russian).

Kitaev, A.B., & Rochev, A.V. (2008). Hydrochemical regime of the dam part of the Kama reservoir. *Geographic Bulletin*, 2, 123–138. (in Russian).

Klimchouk, A., Palmer, A.N., Waele, J.De, Auler, A.S., Audra, P. (Eds.). (2017). *Hypogene karst regions and caves of the world*. Springer.

Kovaleva, T.G. (2010). Types of geological sections in the region of the Ust-Kishert village and karst occurrences). *Geology and mineral resoureces of the Western Urals*, 10, 260–263. (in Russian).

Kovaleva, T.G. (2011). Karst manifestations of the Suksunsky region of the Perm Region. *Geology and mineral resoureces of the Western Urals*, 11, 155–157. (in Russian).

Kovaleva, T.G., & Nikolaev, P.V. (2013). Karst hazard of the territory of the Suksun village of Perm Region. *Modern Directions of Theoretical and Applied Researches*, 1(46), 51–54. (in Russian)

Kostarev, S.M. (2004). Formation of technogenic accumulations of components of deep fluids in near-surface rock massifs (by the example of oil production areas in the Perm region. *Proceedings of higher educational institutions. Oil and gas*, 5, 132–143. (in Russian).

Kostarev, S.M. (2015). Features of oilfield pollution of the geological environment of the karst regions of the Perm Region. *Environmental safety and construction in karst regions*, 317–322. http://nsi.psu.ru/cave/bibl/karst2015.pdf. (in Russian).

Kostarev, V.P., Dimukhametov, M.Sh., Malakhov, V.E., Speshkov, L. N. (1986). Activation of karst manifestations during the construction of gas trunklines. *Geological environment and rational use of mineral resources of the Perm region.* (in Russian)

Krutik, I.A., & Kilin, Yu.A. (2018). Modern chemical composition of the waters of the lakes of the Mazuevskaya karst depression. *Bulletin of the Perm University. Geology*, 17(3), 245–251. (in Russian).

Krutik, I.A., & Ovsyannikova, E.S. (2018). Modern chemical composition of lake waters of the Dikoozerskaya karst depression. *Geology and mineral resources of the Western Urals*, 1(38), 232–235. (in Russian)

Kudryashov, A.I. (2013). *Verkhnekamskoe salt deposit*. Moscow: Publishing house "EPSILONPLUS". (in Russian).

Kuznetsov, A.M. (1952). The fibrous gypsum in the Kungur strata of the Urals. Proceedings of the Natural Science Institute at the Perm State University, 13(6), 503–518. (in Russian).

Kungur Ice Cave: Experience of Regime Observations. (2005) Dublyansky, V.N. (ed.). Ekaterinburg: Ural Branch of the Russian Academy of Sciences. (in Russian).

Lavinskaya, O.V., & Khlevnyuk, O.V. (2008). *Prisoners at the construction sites of communism. GULAG and energy facilities in the USSR.* (in Russian)

Lavrov, I.A. (1992). Caves of the Mazuevskaya karst depression. *Study of the Ural caves: reports of the 2nd and 3rd conferences of Ural speleologists*, 49–53. (in Russian)

Lavrova, N.V. (2005). *Regularities of distribution and formation of karst breccias (on the example of the Perm Kama region): abstracts of PhD thesis.* (in Russian).

Lavrova, N.V. (2013). Comparative characteristics of gypsum deposits in the Kungur region of the Perm Region. *Strategy and processes of development of georesources*, 11, 42–44. (in Russian).

Lavrova, N.V. (2018). The karst nature of the Cis-Ural depression. *Geology, geography and global energy*, 1, 43–47. (in Russian).

Liu, Yu. (2012). *Modeling the influence of salt dumps of potash plants on the filtration properties of water-bearing rocks (on the example of the territory of influence of the 2nd Solikamsk potash mine administration): abstracts of PhD thesis.* (in Russian).

Liu, Yu., & Lekhov, A.V. (2012). Modeling of changes in filtration parameters of gypsum rocks during brine filtration. *Geoecology. Engineering geology. Hydrogeology. Geocryology*, 6, 551–559. (in Russian).

Lukin, V.S. (1964). Sinkhole phenomena in the Urals and in the Cis-Urals. *Hydrogeological collection*, 3, 133–160. (in Russian).

Lukin, A.V., & Ezhov, Y.U. (1980). Large-scale engineering-geological zoning of the territory of the Krasny Yasyl village of Perm region. *Karst of the Non-Black Earth Region*, 103–104. (in Russian).

Lyagushkin, V., Vaschenko, B., Maksimovich, N., Lavrov, I., Pankov, N., Shumeiko, I., Klimchuk, A., Runkov, E. (2011). *Ordinskaya cave* (an illustrated collection of articles). Moscow: Studio "4 + 4". (in Russian).

Lyadova, N.A., Yakovlev, Yu.A., Raspopov, A.V. (2010). *Geology and development of oil fields in the Perm Region.* Moscow: OAO VNIIOENG. (in Russian)

Maklashin, A.V., & Alvanyan, A.K. (2015). Engineering-geological and geophysical studies to identify karst-hazardous sites on the Sverdlovsk railway. *Environmental safety and construction in karst regions*, 107–109. http://nsi.psu.ru/cave/bibl/karst2015.pdf. (in Russian).

Maksimov, A.Yu., Pozyumko, E.N., Pavlova, Yu.A., Kataev, V.N. (2017). Metagenomic analysis of microflora of karst lakes in the Kishertsky karst region. *History and methodology of physiological, biochemical and soil research*, 93–94. (in Russian).

Maksimovich, G.A. (1947). Speleographic sketch of the Molotov region. *Speleological bulletin of the Natural-Scientific Institute at the M. Gorky Molotov State University*, 5–42. (in Russian).

Maksimovich, G.A. (1964). Earthquakes in the Kama region. *Geography of the Perm region*, 2, 133–140. (in Russian).

Maksimovich, G.A. (1969). *Fundamentals of karst studies. V. 2. Issues of hydrogeology of karst, rivers and lakes of karst regions, chalk karst, hydrothermokarst.* Perm. un-t (Geographical Society of the USSR. Institute of Karst Studies and Speleology). (in Russian)

Maksimovich, N.G. (2006). *Safety of dams on soluble rocks (on the example of the Kama hydroelectric power station).* Perm: OOO PS "Garmoniya". http://nsi.psu.ru/labs/gtp/stat/2006/0298.pdf. (in Russian).

Maksimovich, N.G. (2008). Mineralogy of the Ordinskaya cave. *Problems of mineralogy, petrography and metallogeny*, 11, 72–77. http://nsi.psu.ru/labs/gtp/stat/2008/0332.pdf. (in Russian).

Maksimovich, G.A., Chistyakov, N.M. (1957). Healing mud of the Suksunsky district of the Perm Region. *Scientific notes of the Molotov University*, 11(2), 55–58. (in Russian).

Maksimovich, G.A., & Gorbunova, K.A. (1958). *Karst of the Perm region*. Perm: Perm publishing house. http://nsi.psu.ru/cave/bibl/kpo.pdf. (in Russian).

Maksimovich, N.G., & Bolotov, G.B. (1982). Features of the karst relief of the Polazninsky site. *Physical and geographical bases of development and distribution of productive forces of the Non-Black Earth Urals*, 84–87. http://nsi.psu.ru/labs/gtp/stat/ng_0037.html. (in Russian).

Maksimovich, N.G., & Khmurchik, V.T. (2006). *Sampler*. Patent 54398 Russian Federation. (in Russian)

Maksimovich, N.G., & Khmurchik, V.T. (2007). *Consortium of strains of hydrocarbon-oxidizing bacteria pseudomonas aeruginosa nd kz-1 and pseudomonas fluorescens nd kz-2 as a destructor of oil products and a method for purifying oil-contaminated groundwater.* Patent 2312719 Russian Federation. (in Russian).

Maksimovich, N.G., Meshcheriakova, O.Yu. (2009a). The mechanism of oil pollution in the area of karst banks of reservoirs and methods for improving the ecological situation. *Modern problems of reservoirs and their watersheds.* Vol. 1: Hydro- and geodynamic processes. Chemical composition and quality of water, 265–270. http://nsi.psu.ru/labs/gtp/stat/2009/0350.pdf. (in Russian).

Maksimovich, N.G., Meshcheriakova, O.Yu. (2009b). Methods of dealing with oil pollution on the karst shores of reservoirs. *Ecology of urbanized territories*, 4, 55–58. http://nsi.psu.ru/labs/gtp/stat/2009/0359.pdf. (in Russian).

Maksimovich, N.G., Meshcheriakova, O.Yu. (2009c). The influence of gypsum karst on hydrotechnical constructions in Perm region. Geological Engineering Problems in Major Construction Projects, 2, 604–607.

Maksimovich, N.G., & Khayrulina, E.A. (2012a). Fundamentals of environmental monitoring during the development of a deposit of potassium salts. *Engineering research*, 8, 20–30. http://nsi.psu.ru/labs/gtp/stat/2012/0397.pdf. (in Russian).

Maksimovich, N.G., & Pervova, M.S. (2012b). Influence of overflows of saline waters of the Verkhnekamskoye potassium-magnesium salt deposit on the near-surface hydrosphere. *Engineering researches*, 1, 22–28. http://nsi.psu.ru/labs/gtp/stat/2012/387.pdf. (in Russian).

Maksimovich, N.G., & Pyankov, S.V. (2012c). *Small reservoirs: ecology and safety*. Perm: publishing house "Raritet - Perm". http://nsi.psu.ru/labs/gtp/stat/2012/0393.pdf. (in Russian).

Maksimovich, N.G., Meshcheriakova, O.Yu. (2017). Development of Sulphate Karst Under Technogenic Impact Conditions in the Western Urals. EuroKarst 2016, Neuchtel. *Advances in the Hydrogeology of Karst and Carbonate Reservoirs*, 65–70.

Maksimovich, N.G., Meshcheriakova, O.Yu. (2018a). The Experience of the Dam Construction on Gypsum-Bearing Rocks (in the Territory of the Former USSR). *Karst 2018. Expect the Unexpected*, 81–88.

Maksimovich, N.G., & Pyankov, S.V. (2018b). *Kizel coal basin: ecological problems and solutions*. Perm: Perm State National Research University. http://nsi.psu.ru/labs/gtp/stat/2018/529.pdf. (in Russian).

Maksimovich, N.G., & Sergeev, V.I. (1983). Effect of chemical injection stabilization on gypsum stability in the foundation of hydraulic structures. *Hydrotechnical construction*, 17(7), 380–384.

Maksimovich, N.G., Maksimovich, E.G., Lavrov, I.A. (2006). Ordinskaya cave: the longest underwater cave in Russia. Perm. http://nsi.psu.ru/labs/gtp/stat/2006/0320.pdf. (in Russian).

Maksimovich, N.G., Khmurchik, V.T., Meshcheriakova, O.Yu. (2016a). Bacterial sulfate-reducing process in oil-polluted karst rocks. *Goldschmidt Conference Abstracts*. Yokohama. http://goldschmidt.info/2016/uploads/abstracts/proofs/M.pdf.

Maksimovich, N.G., Meshcheriakova, O.Yu., Kachurin, N.M. (2016b). Cleaning of gypsum deposits from oil pollution in karst regions. *Gornyi Zhurnal (Mining Journal)*, 2, 92–97. http://nsi.psu.ru/labs/gtp/stat/2017/500.pdf. (in Russian).

Maksimovich, N.G., Milanović, P., Meshcheriakova, O.Yu. (2017). Dams on Evaporite Rocks Foundation. *Man and Karst 2017*, 37–50.

Maksimovich, N., Meshcheriakova, O., Khmurchik, V. (2019). Bacterial Processes in Oil-Polluted Karst Environments in Perm Region (Russian Federation). *IAEG/AEG Annual Meeting Proceedings*, San Francisco, California, 3, 103–107.

Malakhova, T.E., Minkevich, I.I., Papirovaya, V.T. (2004). Hydrogeological conditions for the development of karst in the Demidkovsky tract (Polazninsky karst region). *Karst studies - XXI century: theoretical and practical significance*, 187–191. http://nsi.psu.ru/cave/bibl/karstvXXI.pdf. (in Russian).

Mamenko, G.K. (1969). Studying the safety of gypsum-bearing rocks at the base of the Kama hydroelectric power station). *Hydrotechnical construction*, 5, 16–21. (in Russian).

Matarzin, Yu.M. (2003). *Hydrology of reservoirs*. Perm: PSU, PSI, PSSGK. (in Russian).

Maximovich, N.G., Kazakevich, S.V., Khmurchik, V.T. (2005). Development of methods protection of the Kama reservoir from oil pollution. *Quality and management of water resources*, 362–369.

Maximovich, N., & Shumilova, O. (2008). Ordinkaya Cave – The Longest Underwater Cave in Russia. *3-rd International Workshop on ice Caves*, 105–107.

Meshcheriakova, O.Yu. (2014). *Protection of gypsum deposits from oil pollution in the karst regions of the Perm Region: PhD thesis*. (in Russian).

Mikhailov, G.K, & Oborin, A.A. (2006). *Underground storage room of the Sylvensky ridge*. Perm: Ural Branch of the Russian Academy of Sciences; Perm State Universuty. (in Russian).

Milanović, P., Maksimovich, N., Meshcheriakova, O. (2019a). *Dams and Reservoirs in Evaporites*. Springer, Cham.

Milanović, P., Maksimovich, N., Meshcheriakova, O. (2019b). Experience with some investigation and remediation methods in evaporite karst. *Geotechnical Challenges in Karst. Karl Terzaghi and karst in Croatia 110 years ago*, 281–286.

Minkevich, I.I., & Kilin, Yu.A. (2018). Metamorphization of the chemical composition of groundwater in the karst regions of the Kama region. *Geological evolution of the interaction of water with rocks*, 131–135. (in Russian).

Minkevich, I.I., Kilin, Yu.A., Kashevarova, A.A., Oshchepkov, V.I. (2019). Ecological and hydrogeological characteristics of the Chusovsky cape of the Kamskoye reservoir. *Sergeevskie readings. Ecological and economic balance of nature management in mining regions*, 21, 454–459. (in Russian).

Minkevich, I.I., & Oshchepkov, A.A. (2004). Karst lakes of the Perm Region (Polazninsky karst region). *Karst studies – XXI century: theoretical and practical significance*, 184–187. http://nsi.psu.ru/cave/bibl/karstvXXI.pdf. (in Russian).

Naugolnykh, S.V., & Zavjalova, N.E. (2004). Densoisporites polaznaensis sp. nov.: with comments on its relation to Viatcheslavia vorcutensis Zalessky. *Paleobotanist*, 53, 21–33.

Nazarov, N.N. (1996). *Karst of the Kama region. Physico-geographical (geomorphological) aspects*. Perm. (in Russian).

Nazarov, N.N. (2004). Karst shores of the Kama reservoir: distribution, processing intensity, classification. *Karst studies – XXI century: theoretical and practical value*, 122–130. http://nsi.psu.ru/cave/bibl/karstvXXI.pdf. (in Russian).

Nazarov, N.N., Frolova, I.V. (2009). Features of the morpholithogenesis of the karst shores of the Kama reservoir. *Geomorphology*, 4, 92–98. (in Russian)

Oborin, A.A., Khmurchik, V.T., Ilarionov, S.A., Markarova, M.Yu., Nazarov, A.V. (2008). *Oil-contaminated biogeocenoses (Formation processes, scientific foundations of restoration, medical and ecological problems)*. Perm: Ural Branch of the Russian Academy of Sciences; Perm State University; Perm State Technology University. (in Russian).

Oradovskaya, A.E. (1962). Discharfe of saline waters as a factor in protecting the foundation of a dam from dissolution. *Scientific communications of VODGEO. Hydrogeology*, 26–34. (in Russian).

Osipov, V.I., Baryakh, A.A., Sanfirov, I.A., Mamaev, Yu.A., Yastrebov, A.A. (2014a). Karst hazard during flooding of a potash mine in the Berezniki town of Perm Region of the Russian Federation. *Geoecology. Engineering geology, hydrogeology, geocryology*, 4, 356–361. (in Russian).

Osipov, V.I., Zverev, V.P., Kostikova, I.A., Mamaev, Yu.A. (2014b). The hydrogeochemical features of the interaction of water and rocks in the Verkhnekamsky salt-bearing basin). *Geoecology. Engineering geology, hydrogeology, geocryology*, 5, 402–407. (in Russian).

Pecherkin, I.A. (1969). *Geodynamics of the coasts of the Kama reservoirs*. Part 2: Geological processes. Perm. (in Russian).

Pecherkina, L.V. (1983). *Hydrogeology and hydrogeochemistry of the Polaznensky gypsum-anhydrite massif*. Perm: Perm State University; VNITI. (in Russian)

Pecherkin, A.I. (1986). *Geodynamics of sulfate karst*. Irkutsk: Irkutsk University. (in Russian).

Pecherkin, A.I., & Zakoptelov, V.E. (1982). *Karst and suffusion on the banks of reservoirs*. Perm: Perm University. (in Russian).

Pecherkin, I.A., Karzenkov, G.I., Matskevich, I.K. (1965). Ephemeral caves. *Peshchery (Caves)*, 5(6), 86. http://nsi.psu.ru/cave/vipuski/v5(6).pdf. (in Russian).

Pecherkin, A.I., Maksimovich, N.G., Zakoptelov, V.E. (1980). Types of cave on Kama reservoir scores. *Europ. regional conf. of speleology*, 34.

Pecherkin, A.I., Maksimovich, N.G., Bolotov, G.B., Zakoptelov, V.E. (1981). Caves and other karst forms in gypsum and anhydrite on the coast of the Kama reservoirs. *Peshchery (Caves)*, 18, 55–58. http://nsi.psu.ru/labs/gtp/stat/ng_0033.html. (in Russian).

Pikovsky, Yu.I., Smirnova, M.A., Kovach, R.G., Puzanova, T.A., Khlynina, A.V., Khlynina, N.I. (2016). Bituminous substances and polycyclic aromatic hydrocarbons in soils above the lenses of oil and oil products in underground karst cavities (Polaznensky Peninsula, Perm Region). *Soil studies*, 3, 321–332. (in Russian)

Popov, L.N., & Maksimovich, N.G. (2008). *Installation for pumping oily liquid from a well*. Patent 81522 Russian Federation. (in Russian).

Potapov, A.A. (2016). Regional numerical geofiltration model of the southern part of the Verkhnekamskoe potassium-magnesium salt deposit). *Bulleten of SPbGU. Geology. Geography*, 3, 4–23. (in Russian).

Royer, J.J., Litaudon, J., Filippov, L.O., Lyubimova, T., Maximovich, N. (2017). 3D geostatistical modelling for identifying sinkhole disaster potential zones around the Verkhnekamskoye potash deposit (Russia). *Journal of Physics: Conference Series*, 879(1), 012018.

Shimanovsky, L.A. (1973). *Fresh underground waters of the Perm Region*. Perm. (in Russian).

Shimanovsky, L.A., & Kuntz, E.V. (1986). Formation of the chemical composition of groundwater in the zone of active water exchange in the Perm Region. Chemical composition and groundwater resources of the Cis-Urals and Trans-Urals, 19–33. (in Russian)

Shumilova, O.Yu., & Maksimovich, N.G. (2008). Distribution of karst in the administrative regions of the Perm Region. *Problems and tasks of engineering and construction surveys. Problems of engineering geology of karst in urbanized territories and reservoirs*, 294–301. http://nsi.psu.ru/labs/gtp/stat/2008/0335.pdf. (in Russian).

Snitko, G.P., Gai, V.V., Dresvyannikova, N.F. et al. (2017). *State Geological Map of the Russian Federation*. Scale 1:200,000. Perm series. Sheet O-40-XVI (Lysva). Explanatory letter. Moscow: MF FGBU "VSEGEI". (in Russian).

Snitko, G.P., Popova, T.N., Gorbunova, M.K. et al. (2015). *State geological map of the Russian Federation*. Scale 1:200,000. Second edition. Perm series. Sheet P-40-XXVIII (Colva). Explanatory letter. Moscow: MF VSEGEI. (in Russian).

Specially protected territories of the Perm Region: register. (2002). Ovesnov, S.A. (ed.) Perm: "Knizhnyy mir". (in Russian).

Specially protected natural territories of the Perm Region. (2017). Buzmakov, S.A. (chief editor). Perm: Aster. (in Russian).

Tchaikovsky, I.I. (2008). The main stages of the formation of tectonic faults in the water-protective strata of the Verkhnekamskoye potassium deposit). *Gornyy zhurnal (Mining Journal)*, 10, 41–44. (in Russian).

Tchaikovsky, I.I. (2018). A new genetic type of brines from the Verkhnekamskoye deposit. *Strategy and processes of development of georesources*, 3–7. (in Russian).

Turyshev, A.V. (1960). The formation of depressions along the slopes of the Ufa plateau. *Issues of geomorphology and newest tectonics of the Volga-Ural region and the South Urals*, 205–214. (in Russian).

Tyurina, I.M., Minkevich, I.I., Ponomarev, A.B. (1994). Conditions for the development of a karst sinkhole on the Chusovskaya spit of the Kama reservoir. *Karst sinkholes*, 86–87. (in Russian).

Utemova, S.A., & Papirovaya, V.T. (2015). The karst hazard of the Perm urban agglomeration. *Environmental safety and construction in karst regions*, 205–209. http://nsi.psu.ru/cave/bibl/karst2015.pdf. (in Russian).

Valuisky, S.V., Rodionov, V.V., Belokrys, I.A. (1988). The longest and deepest caves of the Perm region. *Peshchery (Caves)*, 21, 123–127. http://nsi.psu.ru/cave/vipuski/v21.pdf. (in Russian).

Varsanofieva, V.A. (1915). Karst phenomena in the northern part of the Ufa plateau. *Earth science*, 21(4), 39–83. (in Russian).

Vodolazskaya, V.P., Teterin, I.P., Kirillov, V.A., Lukyanova, L.I. et al. (2015). *State geological map of the Russian Federation*. Scale 1:1,000,000 (third edition). Ural series. Sheet O-40. Perm. Explanatory letter. SPb.: Cartographic factory VSEGEI. (in Russian).

Voronkevich, S.D., Evdokimova, L.A., Sergeyev, V.I. (1976). Theoretical foundations and results of the introduction of a method for chemical plugging of semi-rocks and rocks. *Issues of engineering geology and soil science*, 4, 199–209. (in Russian).

Yakovenko, P.I. (1968). Development of karst processes in the Middle Cis-Urals. *Design, construction and operation of the roadbed in karst regions*, 180–190. (in Russian).

Yarkeeva, N.R. (1999). Estimation of the equilibrium saturation of produced waters with calcium sulfate depending on their total mineralization. *Development and operation of oil fields*, 168–174. (in Russian).

Yashchenko, R.V. (1984). Karst river Turaevka as a natural monument of the Perm Region. *Methods of geological research*, 68–69. (in Russian).

Zaderigolova, M.M., Zaderigolova, M.M. (Jr.), Bilak, E.V. (2015). The significance of the radio wave method in assessing karst hazard and risk. *Environmental safety and construction in karst areas*, 163–166. http://nsi.psu.ru/cave/bibl/karst2015.pdf. (in Russian).

Zverev, V.P. (1967). *Hydrochemical studies of the gypsum-groundwater system*. Moscow: Publishing house "Nauka". (in Russian).